KU-521-980

Contents

	Introduction	iv
1	Brass or Bronze?	1
2	Casting Copper	13
3	Red Brass	43
4	Leaded Yellow Brass	67
5	High-Strength Yellow Brass	74
6	Tin and Leaded-Tin Bronzes	119
7	High-Leaded Tin Bronze	131
8	Silicon Bronze and Silicon Brass	150
9	Phosphor Bronze	163
10	Aluminum Bronze	167
11	Other Red-Metal Alloys	181
12	Commonsense Gates and Risers	189
13	Revelations	211
14	The Business	229
	Appendix: Sand Mixes	240
	Index	244

Introduction

This is not just another metal-casting text. It would be a simple matter for me to fill a volume with a pile of information on brass-casting sands, core mixes, and gating. While I will cover this material to the extent that I think it is necessary (plus some special conditions and processes), the purpose of this book is to explain how to produce a quality brass or bronze casting and also to describe the workings of the brass foundry.

Knowing how to come up with a top-quality casting is only the tip of the iceberg. There is much more to it than that. Like the horns of the bull, one facet is casting ability and the other is operating a viable casting business. Between the horns lies the dilemma. It is possible to set up a small, medium, or even a very large foundry—brass, aluminum, cast iron or whatever—and go down the tubes knowing full well how to produce good, salable castings, but not knowing how to operate the business end of the foundry. You must know when to send a character and his pattern packing and when not to buy that bigger compressor, etc. The misunderstanding of these business factors is more likely to bust you than shoddy castings.

It's not like it used to be. A couple of fellows could set up and kick off a small foundry and make a nice living at it without much hassle. It's a new ball game. You can operate a small one- or two-man brass foundry today and be more efficient at what you do than the larger shops. It is a good trick but not hard to learn.

Keep things small and own it all is the first rule. The second rule is don't get greedy. Poor management is the downfall of many a business—as is too much management. We have two extremes in which we can get into with a horse. First, we can all be in front feeding the horse with nobody around back with a shovel taking care of that end. It doesn't take much imagination to see what would happen (that's where the country is today). Second, we could all

CASTING BRASS

C.W. AMMEN

TAB BOOKS Inc.
Blue Ridge Summit, PA 17214

To My Wife, Hazel

Other TAB Books by the Author

No. 725 *Lost Wax Investment Casting*
No. 1043 *The Complete Handbook of Sand Casting*
No. 1173 *The Metalcaster's Bible*
No. 1510 *Constructing and Using Wood Patterns*
No. 1610 *Casting Iron*
No. 1910 *Casting Aluminum*

N21 554 1809

FIRST EDITION

FIRST PRINTING

Copyright © 1985 by TAB BOOKS Inc.

Printed in the United States of America

Library of Congress Cataloging in Publication Data

Ammen, C. W.
    Casting brass.

    Includes index.
    1. Brass founding.   2. Bronze founding.   I. Title.
TS565.A46   1985   673 .3   85-4676
ISBN 0-8306-0810-9
ISBN 0-8306-1810-4 (pbk.)

be around back of the horse with shovels with no one around front feeding the horse.

The proper ratio between the fellows up front and the fellows behind the horse is what we need for a viable, profitable operation. With the one-man operation, you have to run from end to end. That's a mean task, but at least you have control of both ends.

The time was when you had little or no problem getting foundry supplies in small quantities, and the foundry supply houses were owned and operated by actual foundry experts. Today most, if not all, foundry suppliers operate on a large scale and are not interested in the small operator who wants to purchase 10 pounds of dry parting, a couple of #20 crucibles, etc.

Their answer is that they lose money that way, and this could be true. Nevertheless, where does that leave the small-time operator? You are going to have to improvise quite a bit to make a go of it on a small scale. You can build your own melting equipment, find your own sand, and use sill sand for parting. You need core oil? The supplier will sell you a 55-gallon drum. Instead, use linseed oil sold at the local lumberyard or paint-supply store. Need molasses? Buy a bottle of molasses in the supermarket.

Have you decided on a small rotary furnace as your melting unit? You can buy one for $20,000 or $30,000, plus freight. Fine, so you build one. For a burner and blower, the price is out of this world. Build a burner out of pipe fittings and blow it with a shop vac or shop the junkyards for an old blower.

A Canadian friend of mine makes his own crucibles. I have never tried this. Crucibles are extremely expensive. If you can, your best bet is a small rotary for brass and a small little reverb for aluminum. Ladles can be fabricated, and with a little fire clay loam you can line them.

You used to be able to walk in a local foundry and look up the gaffer and bum some parting, etc. This is impossible today. You can't get past the office. Even used foundry equipment is usually out of reach of the backyard foundryman. With an engine lathe, cutting torch, and a small electric welder, you can fabricate about everything you need.

Going from the ground up, you could locate some natural-bonded molding sand, build a small cupola out of a couple of 55-gallon drums, make a few wood flasks, make the patterns you need, and cast the parts you need to build a rotary furnace and flask hardware. Simply cast yourself a foundry. Melt up old automobile blocks and such.

There is nothing more basic than green-sand molding. A lot of foundries that have moved away from green sand to the more so-called high-tech processes (no-bake, vacuum molding, etc., etc.,) are presently re-evaluating and moving back to green sand to cut costs and become more competitive again. You must think low technology.

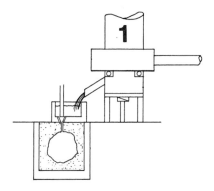

Brass or Bronze?

Brass and bronze are commonly referred to as red metals or nonferrous metals. I prefer the term red metals as the term nonferrous implies any metal other than iron or steel, which covers one heck of a lot of other metals. Some define nonferrous as containing no iron content whatsoever. Medium tensile manganese bronze is composed of 64 percent Cu, 26 percent Sn, 3 percent iron, 4 percent aluminum and 3 percent manganese.

The term brass foundry is the common term for a foundry that produces copper-base alloys of brass and bronze. Many brass foundries might cast a small amount of zinc-base alloys and some aluminum. If, however, the main endeavor is copper-base alloys (red metals) it is called a brass foundry (never a bronze foundry). The term bronze works is usually a foundry engaged in the production of bronze bushings, propellors, and the like.

Pure copper is defined as not having over 2 percent total content of lead, tin, zinc, arsenic, cadmium, silicon, silver etc., elements. To be classified as copper, it must be 98 percent pure copper.

When copper is alloyed past this 2-percent figure, it becomes a brass or bronze-copper base alloy. From here things become very confusing. Let's look at manganese bronze. Manganese bronze is actually not a bronze at all; it is actually a high-strength yellow brass. A medium-strength manganese bronze would be an alloy like this: 64 percent copper, 26 percent zinc, 3 percent iron, 4 percent

Table 1-1. Copper-Based Alloys.

A = An alloy not commonly used in the foundry for casting
B = A common foundry casting alloy

Red brass (A)
Zinc 2 to 8%
Tin less than zinc
Lead less than .5%
Copper remainder

Leaded red brass (B)
Zinc 2 to 8%
Tin less than 6%
Lead over .5%
Copper remainder

Semi-red brass (A)
Zinc 8 to 17%
Tin less than 6%
Lead less than .5%
Copper remainder

Leaded semi-red brass (B)
Zinc 8 to 17%
Tin less than 6%
Lead over .5%
Copper remainder

Yellow brass (B)
Zinc over 17%
Tin less than 6%
Lead less than .5%
Copper remainder

Leaded yellow brass (B)
Zinc over 17%
Tin less than 6%
Lead over 5%
Copper remainder

High-strength yellow brass (B)
Zinc over 17%
Aluminum-Manganese-Tin-Nickel-Iron (over 2% total)
Silicon less than .5%
Lead less than 6%
Tin less than 6%
Copper remainder

Leaded high-strength yellow brass (B)
Zinc over 17%
Aluminum-Manganese-Tin-Nickel-Iron (over 2% total)
Lead over .5%
Tin less than 6%
Copper remainder

Silicon brass (B)
Any lead free brass with

Tin brass (A)
Tin over 6%

a silicon content over .5% With zinc over 3% Copper remainder	**Zinc more than tin** Copper remainder
Tin nickel brass (A) Tin over 6% Nickel over 4% Zinc more than tin Copper remainder	**Nickel brass** (Nickel Silver) (B) Enough nickel to give the alloy a white color. Lead under .5% Copper remainder
Leaded nickel brass (leaded nickel silver) (B) Zinc over 10% Enough nickel to give the alloy a white color. Lead over .5% Copper remainder	
Tin bronze Tin 2 to 20% Zinc less than tin Lead less than .5% Copper remainder	**Leaded tin bronze** (B) Tin 2 to 20% Tin 2 to 20% Lead over 5% but under 6% Copper remainder
High lead tin bronze (B) Tin 2 to 20% Zinc less than tin Lead over 6% Copper remainder	**Lead bronze** (B) Lead over 30% Tin less than 10% Zinc less than tin Copper remainder
Leaded nickel bronze (B) Nickel over 10% Zinc less than nickel Tin less than 10% Copper remainder	**Aluminum bronze** (B) Aluminum 5 to 15% Iron up to 10% Lead less than .5% Copper remainder Note: aluminum bronzes usually contain some manganese and nickel.
Silicon bronze Silicon over .5% Zinc not over 3% Copper not over 98%	**Beryllium bronze** (B) Beryllium over 2% Copper remainder

aluminum, 3 percent manganese, tensile strength 95,000 PSI.

Now this is not a bronze but a yellow brass, as it is a very high-strength, durable casting alloy. It would never do to call it a yellow brass, which is considered as a cheap yellow casting metal used for novelties and castings that do not require strength. A typical yellow brass would look like this: 64 percent copper, 1 percent tin, 3 percent lead, 29 percent zinc, tensile strength 12,000 PSI.

You see we have two yellow brasses. One is a little over 87 percent stronger than the other so it would never do to call it a yellow brass or a high-strength yellow brass (it contains 3 percent manganese). Let's call it manganese bronze. We call all the beautiful statues and red-metal artwork "Bronze" (sounds more expensive, stronger, etc.) when actually most are red-brass alloys and a great many are cheap yellow brass.

In 1939, the American Society of Testing Materials made an effort to classify copper-base alloys. See Table 1-1. These figures are simply guidelines, and I will get around to the specific alloys and their various correct pedigrees.

After looking over the general guidelines for various brasses and bronzes, you should realize that each branches off into various alloys or subgroups. Let's look at a couple. First, let's look at red brass. Ingots, aside from having an ASTM (American Society for Testing Materials) designation and the smelter's number, are commonly called by the number designating the basic composition. Common red brass is called 85-5-5-5 or 85 three fives-ounce metal, or ingot #15 etc., etc.

Let's look at them by composition numbers. The 85-5-5-5 would simply mean that this alloy consists of 85 percent copper (Cu), 5 percent tin (Sn), 5 percent lead (Pb), and 5 percent zinc (Zn).

The composition is always stated copper first, tin next, lead next, and then the zinc. The 85-5-5-5—according to the ASTM classifications—is an alloy we would call a leaded red brass. There is less than 6 percent tin and over .5 percent lead, and the zinc—being 5 percent—falls within the tolerance of 2 percent to 8 percent zinc for a red brass.

Suppose you had ingot #123 or 81-3-7-9 (81% Cu-3% Sn-7% Pb-9% Zn). This is a leaded, semi-red brass. There are five common foundry alloys that can be called red brasses: 85-5-5-5; 83-4-6-7; 81-3-7-9; 78-3-7-12; 76-3-6-15.

As the zinc increases, the color becomes lighter and lighter and when we cross over to about about 19 percent zinc we are off into the yellow brasses: 73-1-3-24 = high copper yellow brass; 67-1-3-29

= commercial #1 yellow brass; and 63-1-1-35 = yellow brass (cheap brass).

So we go on and on. If you were to go into a small restaurant and find a multipage menu with everything from hamburgers to roast duck on the menu, it's time to leave. The same applies to a brass foundry. You could hardly ever begin to carry or even claim that you can or will cast virtually every type of brass or bronze available to the trade. So where do you draw the line?

If you are captive, you could simply select the brass or bronze alloy suitable for your finished product and go with that. If you are jobbing in small lots, your best bet is to limit your stock of ingot material to one yellow-brass alloy (commercial #1 yellow 67-1-3-29 and one red brass 85-5-5-5). Here you have two very castable alloys that do not require special gating, treatments, or heavy risers. With both alloys, your practice tolerance are quite broad.

Let's look at 85-5-5-5. This alloy has a tensile strength of 37,000 PSI and a solidification range from 1850° to 1570° F (a range spread out over 280° F). That means you can pour a rather thin casting of considerable length with minimum gating. Pouring temperature for very light castings is 2150° F to 2550° F. Maximum pouring temperature for heavy castings is 1950° F to 2150° F. You will, however, find that if the pouring temperature at the mold is 2150° F to 2200° F, you can pour with good results over a wide range of sizes, weights, and section thicknesses of castings. About 95 percent of anything you get into will fall in this range. So you have an alloy that does not require excessively high melting and pouring temperature (a plus in melting cost and crucible life).

The only treatment required with 85-5-5-5 is one to two ounces of phosphor copper shot per 100 pounds of metal to deoxidize any metallic oxides formed during melting. The procedure is quite simple and the phosphor copper shot is about your least expensive deoxidizer. Super heat your red brass, pull the crucible, skim, and add the shot (just throw it on top). Allow it to do its thing. You will see it working; the metal surface will flatten out and become bright and extremely fluid. One minute after adding the shot, check the temperature and pour.

The shrinkage of 85-5-5-5 is 3/16 of an inch per foot. This gives you an alloy that will not require excessive gating and risering.

The pressure tightness is good and the machinability is excellent (84 on a scale of 100). Of the red brass family, the 85-5-5-5 ingot #115 is in the moderate price range. The color is good and it can be polished and plated easily with a minimum of cost and effort. The castings are useful for a multitude of things from novelties to valves and machine parts.

Now let's look at #1 yellow brass. It is less expensive than red brass, and it is used for novelties, hardware (door knobs, trim), polished or polished and plated.

Lets look at 67-1-3-29 ingot #403. The solidification range is 1725°F to 1700° F (a spread of 25°F). This yellow alloy is very fluid at pouring temperatures for light castings 1900°F to 1050°F, and for heavy castings 1750°F to 1900°F. The tensile strength is 34,000 PSI. The machinability is 80 on a scale of 100. Deoxidizing treatment is not required due to the high zinc content that is continuously purging dissolved oxygen from the molten metal.

One problem with yellow brass is that it is smoky (zinc oxides), as are all high-zinc alloys. Zinc vaporizes at 907°C boiling point (which is just a little above 1650°F). The zinc vapor is the zinc that is vaporizing on the surface of the melt. See Fig. 1-1.

As the zinc fumes come in contact with the air, they will combine with the oxygen to form zinc oxide. Zinc oxide is soluble in acids and alkalies but insoluble in water. It is a nontoxic noncombustible white powder.

I could go on from here and examine everything about zinc, its uses, its compounds, and chemical bonding for hundreds of pages. Any student of the brass foundry should explore each path in great detail.

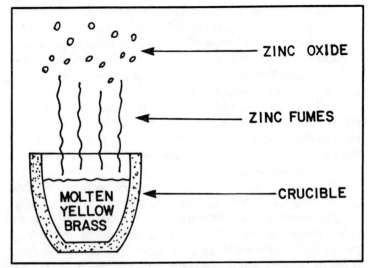

Fig. 1-1 Zinc fumes and zinc oxide rising from the surface of molten yellow brass.

Figure 1-2 shows an ingot made of red brass. Let's look at this drawing as a red brass family tree. From this we see that each constituent of this ingot has its own origins and tree. I will explore with you the ingot as a whole, digressing some but not too far from home base.

With zinc oxide, there are a lot of pros and cons regarding the affect on the human body caused by breathing the zinc oxide present (especially during pouring). I have probably inhaled enough zinc oxide to fill a coal gondola and then some. The only adverse affects in the short run was after a day, where you inhaled a considerable amount of the stuff, you might have what is called the zinc chills that night. Most brass foundrymen knew to drink a half pint to one pint of milk with or after supper and there would be no chills. As to the long-range affects I know of none. Where zinc oxide can really affect you is when the fresh air committee (EPA) sees volumes of this white material exiting your vent stack or window. They come down on you like a drop hammer. The zinc, as a fume, is harmful by inhalation. The tolerance is 5 mg. per cubic meter of air. The zinc oxide as a powder is considered nontoxic.

It stands to reason, however, that too much of anything is too much.

Back to the high zinc problem with yellow brass and other high-zinc alloys, manganese bronze can have from 25 percent Zn to 39.5 percent zinc, and to as high as 40 percent in some die-cast copper alloys. If the gating is not properly designed to fill the mold quickly with a minimum amount of turbulence and the gate system (sprue) is not kept choked (full) during the entire pouring operation, you can get into surface defects on the cope side of the casting (called worm tracks). Defects result if the mold cavity is filled too slowly. The Zn fumes will rise off of the flowing metal and have time to combine with the oxygen present. The zinc oxide that forms will attach itself to the cope surface of the mold. Because zinc oxide has a melting point of 1975 °C, just a little short of 3596 °F, it is simply going to sit there unaffected by heat and cause unsightly defects. See Fig. 1-3.

This type of defect often puzzles the fledgling brass founder because the casting is usually shaken out and wire brushed or blown off—thus destroying the evidence.

If the cope is carefully lifted and the zinc oxide culprit is not disturbed, you will see that tracks in the cope are filled with ZnO powder. See Fig. 1-4.

Now that we have two items on the menu of our brass

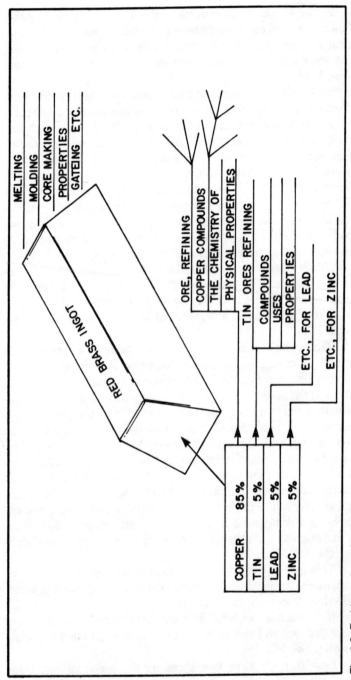

Fig. 1-2. Red brass ingot family tree.

8

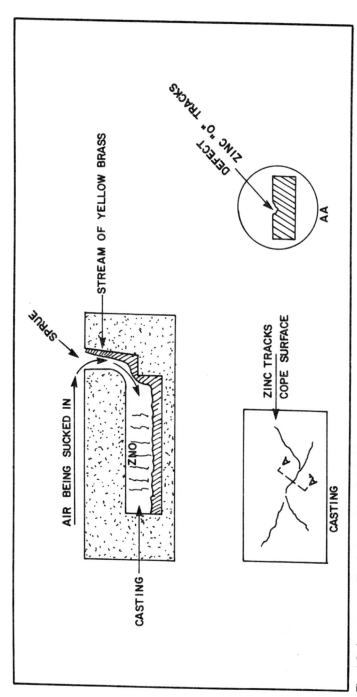

Fig. 1-3. A common yellow brass casting defect (zinc tracks).

9

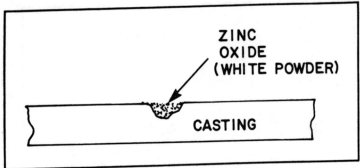

Fig. 1-4. A cross section through the zinc track defect.

foundry, let's add one bronze ingot to give us one red brass, one yellow brass, and one bronze. Bronze is defined very loosely as a copper-tin alloy but the term is used for some alloys containing no tin (such as aluminum bronze). I prefer to go with the tin definition. So which one should you pick? We have hundreds of choices. So let's narrow it down to a few general bronzes and then pick one. Available from 99 percent of the suppliers are:

"G" bronze: 88-8-0-4
Gunmetal: 88-10-0-2
Navy "M": 88-6-1 1/2-4 1/2
Leaded "G": 87-8-1-4
Leaded tin bronze: 87-10-1-2
Leaded tin bronze (no zinc): 88-10-2-0

If you study the uses and typical properties you will soon realize that 88-10-0-2, the gunmetal bronze ingot #210, will cover about 99 percent of all castings requiring a good, high-quality bronze.

With these three ingots, you can produce an extremely wide range of work with a minimum investment and a minimum of foundry problems. After you are at it a while, you will soon establish a ratio of how much you should have of each on hand. For years my colleague Sam and I operated with metal on hand consisting of (by weight) 85 percent red brass (86-5-5-5) 10 percent yellow brass (67-1-3-29) and 5 percent gunmetal (88-10-0-2). Table 1-2 shows the properties of the three chosen for our bill of fare.

Consider the range of major uses we could cover with our menu. The range shown in Table 1-3 is far from complete; there are hundreds of other items. As all three have the same shrinkage (3/16") per foot, this makes the gating and risering similar. The similarity of pouring temperatures means all three alloys can be easily cast in the same system sand.

10

Table 1-2. Brass Properties.

Red Brass: 85% Cu - 5% Sn - 5% Pb - 5% Zn

Properties
Weight Per Cu. In. .318 lbs.
Shrinkage 3/16"
Solidification Range 1850°-1570° F
Pouring Temp. 2100-2350° F light castings
Pouring Temp. 1950-2150° F heavy castings
Tensile strength 35,000 PSI
Machinability 84
Treatment deoxidize with phosphor copper

#1 Yellow: 67% Cu – 1% Sn – 3% Pb – 29% Zn

Properties
Weight per cu. in. .307 lbs.
Shrinkage 3/16" per foot
Solidification Range 1725-1700° F
Pouring Temp. 1900-2100° F light castings
Pouring Temp. 1800-2000° F heavy castings
Tensile strength 38,000 PSI
Machinability 80
Treatment none

Gunmetal bronze: 88% Cu – 10% Sn – 0% Pb – 2% Zn

Properties
Weight per cu. in. .315 lbs.
Shrinkage 3/16" per foot
Solidification Range 1830-1570° F
Pouring Temp. 2100-2300 °F light castings
Pouring Temp. 1920-2100 °F heavy castings
Tensile Strength 45,000 PSI
Machinability 30

Brinell hardness range		Elongation percentage in 2 inches
Red brass	(500 kg) 60	30%
Yellow brass	(500 kg) 35	35%
Gunmetal bronze	(500 kg) 75	25%

The compatibility of these three alloys is the key. You do not have to worry so much about crossbreeding. Should you get a small amount of red brass mixed into a heat of tin bronze, no great harm is done. If a shop tries to cover too large of a selection of alloys to the trade, it becomes a real problem. Many alloys are not compatible, and separate crucibles, ladles, and storage are needed

Table 1-3. Brass Applications.

Red brass:	Valves—Flanges—Pipe fittings—Plumbing goods—Pump castings—Ornaments—Gears—Bushings—Sleeves—Marine hardware—General hardware—Electrical castings—Machine parts glands
Yellow brass:	Knick knacks—Plumbing goods—Ornaments—Name plates—Plaques—Valves—Chandeliers—Ferrules—Bushings
Gunmetal bronze (high grade):	Valves—Fittings—Bearings—Bushings—Piston rings—Gears—Steam castings—Valve slippers—Pumps and impellors—Nuts

before each becomes a mess. With our three basic alloys this is only a minor problem. You can melt and cast a red brass heat, clean out the crucible, and then melt a bronze heat with little chance of crossbreeding or contamination.

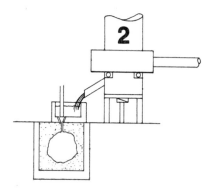

Casting Copper

In Chapter 1, I recommend three basic alloys that I would choose for a small home-brew hobby foundry or small commercial foundry. They might not be your choice so let's take the red metals one by one.

Pure copper castings free from gas porosity are extremely difficult to cast. Your best bet, if you are looking for maximum electrical conductivity (electrical hardware and pole-line castings), is to cast pure copper in dry sand molds (core molds, no-bake furan etc.). Cement molds work well for large, pure copper castings. There is always work for a foundry that can produce high-grade, flaw-free copper castings from extremely small work to large castings (electrode holders, bosh coolers, etc.). The list of items is extremely long; they require high-quality copper castings, brush holders, collector rings, plating tank hardware, special cast bus bars, on and on.

With small copper castings, green sand molds can be used but they must have a high permeability and be low in moisture. Skin drying is very helpful if the mold is not rammed too hard. The weather is a factor in the production of high-quality, gas-free copper castings. If the humidity is high—as on a rainy day or humid day—you will have trouble. Molten copper has a great affinity with oxygen, and on a rainy or humid day, it will absorb or combine with the oxygen present even when the exposure area of the molten copper is only the distance between the pouring lip of the ladle or crucible and the sprue during pouring. In the South, (New Orleans) we would never pour pure copper castings when it was raining or when the humidity was above 60 percent.

Absorption of oxygen during melting with a gas-fired or oil-fired furnace is another factor when the humidity is high. The products of combustion are loaded with H_2O on a rainy or humid day. In Colorado, at 6400 feet above sea level, where a humid day for us is 40 percent, you rarely have a problem. We are talking about relative humidity, the ratio of the pressure of the water vapor actually present in the atmosphere to the pressure of the vapor that would be present if the vapor were saturated at the same temperature. The key word is relative. When you are up here at 6000 feet plus, the pressure is lower and it makes a great deal of difference as compared with sea-level pressure.

Pure copper castings are tough to machine. The metal is stringy and it will roll before the tool. It also is very malleable and requires some doing to get a really smooth, clean-cut surface.

There is a world of published information in regard to exactly how to grind lathe tools and drills. To machine copper, however, I hesitate to give any of this information here as I have found that what seems to work for some is a complete bust for others. I give some general machining tips further down the line. Regardless of what I say or recommend, pure copper is tough to cast, machine, saw, and grind. On a small scale, if you master the founding of pure copper and some limited machining, you should find a lucrative market out there for speciality castings in small lots.

FORM

You can purchase pure copper ingots for melting or you can melt scrap copper. If you have a good source of heavy-high grade copper scrap that is clean and free from iron and solder—and in chunks you can handle—you can go the scrap route. With thin copper scrap (sheet copper, copper tubing and the like), your loss to oxidation will be exceptionally high due to the large surface area. You can cover the material with dry charcoal and reduce this loss, but it is still a hassle (see this chapter's section on melting copper).

If you think you can melt piles of very fine copper wire with any degree of recovery, forget it. Small, thin wire, tubing, and clippings can be briquetted to increase the density against the surface area and melted by ducking these briquettes under the surface of a molten heel of copper. See Fig. 2-1.

You have to watch junk dealers very carefully. The best method

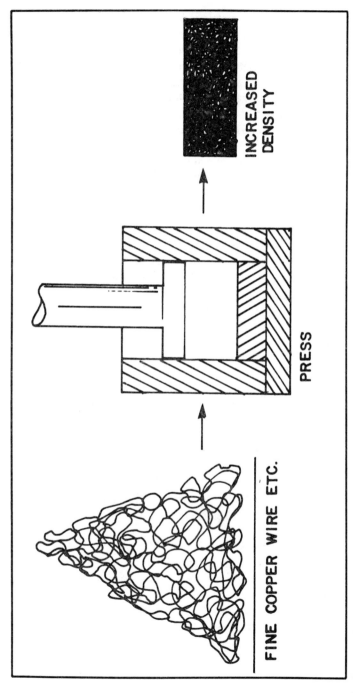

FINE COPPER WIRE ETC.

PRESS

INCREASED DENSITY

Fig. 2-1. Section through a simple briquetting press.

15

for a small operator is to pick and choose. You don't want to buy dirt, paint, lead, and tin. The definition of #1 heavy copper scrap is copper no less than 1/16-inch thick, broken copper castings, heavy field wire, heavy armature wire that is not tangled, new untinned clean punchings and clippings, copper segments that are clean, copper anode butts, heavy bus bars, etc., all of which is clean and free from burnt copper that is brittle and all other foreign substances.

The definition for #1 copper wire is clean, untinned copper wire that is no smaller than 16 gauge on the Brown & Sharp wire gauge. Like #1 heavy, it has to be clean with no foreign substances. If you cannot find scrap copper to fit specifications for your electrical high-conductivity castings, stick with ingot from a reliable metal supplier, and insist on the heat analysis for the batch you purchase.

COPPER MOLDING SANDS

Most natural bonded natural sands that are suitable for light brass are suitable for copper castings. The same holds true for synthetic sands for light brass.

For small, light-green sand molds for copper from 1 ounce to 5 pounds, #"0" Albany sand will do a bang-up job. Moisture should be held at maximum 5.5 percent, green permeability 18, compression strength 4.8 minimum.

A semisynthetic sand (in percent by weight) that will give excellent results is 100 mesh washed and dried silica sand, 47.6 percent, #0, Albany or similar natural bonded sand, 47.6 percent, Southern Bentonite, 4 percent, corn cereal, .8 percent: 47.6 + 47.6 + 4 + .8 = 100 percent.

Now let's not be fussy. I show 100 percent, but you don't have to be this close; these figures are simply guidelines. Having 50 percent silica and 50 percent 0 Albany plus 4 percent bentonite and 1 percent cereal would give you a total of 105 percent and it would be just as good. You have some leeway; just don't stray too far off of base.

This semisynthetic sand tempered with 4 percent moisture should give you a green compression strength of 6.5 to 7, a permeability of 65, and give good results for copper castings up to 50 pounds.

If you like, you can go with a synthetic molding sand of a very simple composition such as 95 percent (lbs) 70 to 100 mesh washed and dried silicia sand, 4 percent (lbs) southern bentonite and 1 percent (lb) wood flour. This would, when tempered with 4.5 percent (lbs) water, give you a green strength of 14.5 and a permeability of 30 plus. Use a good green sand for copper castings up to 10 pounds. If you prefer, the above mix could be 80 percent floor backing sand and 15 percent new sharp silicia in place of 95 percent new sand.

Now let's stop here and consider what we are talking about when we say a particular molding sand is suitable for castings from 1 ounce to 10 pounds. You see this all the time and it can be most confusing. Let's look at two different 10-pound castings (or there about). Let's say we were to cast a solid copper ball 4 inches in diameter. In copper this ball would weigh about 10.7 pounds. Now let's say we were to cast a hollow copper ball with an 1/8-inch wall thickness (section) that weighs 10 pounds. We are looking at a 10-inch diameter ball. See Fig. 2-2.

A molding sand suitable for casting the 12-inch hollow copper ball would not be suitable for casting the solid 4-inch diameter copper ball. The name of the game is more or less a section thickness proposition. Most nonferrous volumes will ramble on with infor-

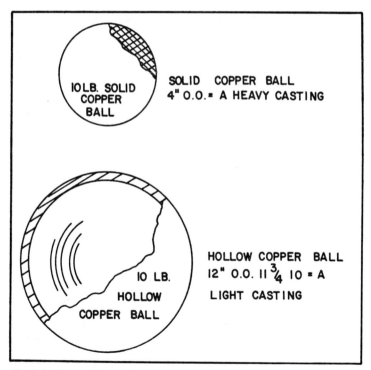

Fig. 2-2. Comparison of a light casting and a heavy casting. Both weigh the same in pounds.

mation as to molding sands and pouring temperatures relating everything to light castings and heavy castings. What are we talking about? The weight of the casting or the section thickness? The average pouring temperature of, for example, regular old red brass (85 percent copper, 5 percent tin, 5 percent lead and 5 percent zinc) would be 2100 °F to 2350 °F for light castings and 1950 °F to 2150 °F for heavy castings.

If you have a casting 3/16 of an inch thick in sections, this would be called a light casting regardless of what the casting itself weighs (1 ounce or 20 pounds). If the casting is 1/2 of an inch thick in section, it's considered a medium-weight casting regardless of its total weight. If it is 1 inch thick (section thickness) it is a heavy casting. Above this, it is a very heavy casting.

You will find that some books will state that all castings weighing over 100 pounds should be cast in dry sand molds or core molds. Other sources put the figure at 300 pounds or over for the breaking point. Gray-iron foundries have been a lot more specific in regard to the definition of what is meant by weight. Even here you will find much confusion. You will find references such as very thin, thin, medium and heavy or thin, medium, heavy or light, very light, medium, medium heavy, heavy, very heavy, thick and very thick and medium thick.

Of course, you have limits as to how far you can run a given metal at its maximum pouring temperature. A casting 3 inches long, and 3/16 of an inch thick in copper is no problem. If the casting is 3/16 of an inch thick and 3 feet long, you have a problem and perhaps an impossible task. See Chapter 12 for examples.

So what we boil this down to is, if we have a flat disc 1 inch thick and 10 inches in diameter, this is a heavy casting. If it is only 3 inches in diameter 1 inch thick, it's still a heavy casting. If it is 1/4 of an inch thick and 10 inches in diameter, it is a light casting. At 3/16 of an inch thick, it is a very light casting. Weight has nothing to do with the reference heavy or light. You can have a heavyweight light casting.

DRY SAND

I have found that with pure copper castings—when you get into a casting that is thick in section regardless of its weight—your best bet is to cast it in a dry sand, core mold, or a no-bake dry mold (furan or silicate of soda).

At any rate, if you use green sand, the mold should be skin dried and poured shortly after drying. If you do not have any way to reclaim sand used to make no-bake molds or core molds (which most small shops do not have) and you do a lot of no-bake molding, it can really cost you in sand. Along with the problem of disposing the unusuable sand, your best bet would be to make dry-sand molds in steel flasks and dry them in the core oven. The sand can then be re-used as heap or backing sand.

Let's clear up the difference between dry-sand molds, core molds, and no-bake molds. Dry-sand molds are green-sand molds to which a binder is added to the molding sand. The mold oven is dried, setting up the binder and driving off the moisture. Common dry-sand mixes suitable for red-metal castings run all over the place. A very good general mix that will work for a large variety of castings consist of 95.5 pounds of dry sand (system sand), 0.2 pounds of corn flour, 3 pounds of bentonite and 1.3 pounds of dextrine. Bake the mold at 350 °F overnight or until dry.

Another good one by volume is 45 measures of 70 or 100 mesh dry silicia sand, 15 measures of natural bonded brass sand (Albany 00), 1 measure of linoil or linseed oil, and 1/2 volume of water. Bake at 425 °F until dry.

Core molds are defined as a mold made of a core-sand mix and baked as you would a core. Any good core mix suited for the metal involved will produce a good core mold from a simple washed and dried silica sand bonded with linseed oil to a more complex mixture.

Mixes suitable for copper core molds are 8 quarts of sharp sand, 1 quart wheat flour, and 1/8 pint of good-grade core oil or boiled linseed oil. Or simply sharp sand and linseed oil: 10 parts oil and 90 parts sand.

The no-bake molds are molds that are made of sharp clay-free sand with a binder which is activated by a catalyst to set the binder (thus gluing the grains of sand together). These molds are called no-bake molds. The same mixes are used to produce molds or cores. No-bake binders are defined as binders that are converted from a liquid to a solid at room temperature.

The term furan is a generic term denoting the basic chemical structure of a class of chemical compounds C_4H_4O, a five-membered heterocyclic compound. The resins produced for no-bake binders are composed basically of furfury alcohol, which is the prime material, in various formulations with ureas and formaldehyde. Actually what you wind up with is a liquid resin that, when catalyzed with acidic materials, forms a tough, resinous film that holds the sand grains together.

The catalyst or activator used is phosphoric acid, 85 percent or 75 percent grade. A typical mix would be washed and dried clay-free sharp silica sand. Binder is 2 percent of the weight of the sand and phosphoric acid is equal to 25 percent of the weight of the binder used. If you have 100 pounds of sand, you need 2 pounds of furan binder and .5 pounds of phosphoric acid.

The amount of catalyst (phosphoric acid) used against the 2 percent binder figure varies depending on how fast a curing time is required and how collapsible you want the core or mold to be.

The procedure for mixing is to mix the catalyst (acid) first with the sand then add the binder. Once mixed, you have to use it immediately because it will be setting up.

If you mixed the binder with the sand first and then added the acid, you will get localized spots of set-up sand, and you would have difficulty in completing the mixing.

If you choose to make large molds with the no-bake method, you can save money by simply using a no-bake mix as a facing sand and filling the remainder of the mold with green sand. The pattern is faced with no-bake mix, and when this mix starts to firm up you can then back it with system sand. You can, if you choose, let the no-bake set up completely then back it with green. See Fig. 2-3.

Because the furan binder will dissolve shellac, if the pattern is shellacked it must be removed with shellac thinner completely. I find that the raw wood (uncoated) pattern gives the best and cleanest parting furan bonded no-bakes. There are various pattern coatings available that are not affected by no-bake bonded sands. Check with your foundry supply house or pattern supply house. They can advise you.

Pure copper castings are very ductile after they solidify in the mold. At the point where it changes from a liqiud to a solid state, however, it is extremely hot short (weak and brittle). If the mold or core does not collapse readily to allow for the casting to shrink, severe hot tears and cracking will occur. In copper casting, this defect is usually caused by the cores and molding material being too high in hot strength. Due to the casting design, the cores for copper castings must be low in hot strength and high in permeability.

The low hot strength allows the core to collapse easily and does not restrict the casting's contraction, thus preventing hot tears. The high permeability and large open vents are required because copper (especially in medium to heavy sections) is extremely susceptible to mold and core gasses. This seems to be due to copper's high conductivity of heat.

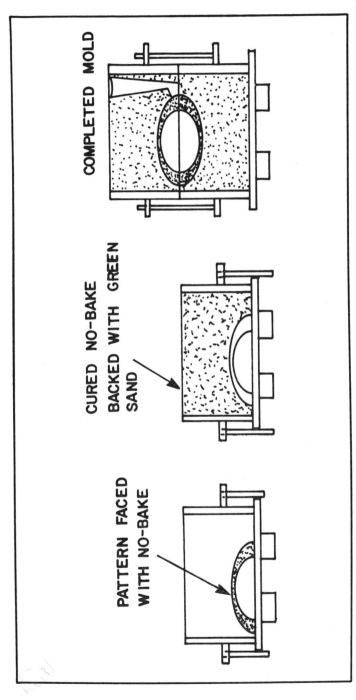

PATTERN FACED
WITH NO-BAKE

CURED NO-BAKE
BACKED WITH GREEN
SAND

COMPLETED MOLD

Fig. 2-3. No-baked facing backed with green sand for economy's sake.

21

This condition prevents the formation of a chill coat or shell at the mold surface and core surface. Therefore, the lack of this gas impervious shell or skin allows the casting to easily absorb any gasses coming from the mold or core surface that is adjacent to the metal itself. If we pour a cast-iron bushing the instant the metal comes in contact with the surface of the mold and the core, a thin layer of metal is solidified, making it impervious to any gas coming from the mold material or core material. This chilled surface comes about by the rapid absorption of the heat in the molten metal that is in contact with the relatively cool mold and core surface. This chill or skin remains solidified and impervious due to the low heat conductivity of the molten iron. Therefore, the molten iron behind this first formed skin allows the skin to remain solidified.

Copper's ability to conduct heat is so great that—as the mold and core surface in contact with the copper tries to chill the copper at its contact surface—it cannot do so because the conductivity of the copper is so great that the contact point remains liquid for a long period.

Let's look at the core and mold material and its gas generation. As the molten metal transfers its heat to the core and mold, the core and mold produce steam and gasses from the water in the sand and any other combustiles and gas-forming materials (such as binders, wood flour, organic matter, sea coal etc., present). The hotter the mold and core gets the greater the gas generation and gas pressure becomes. This gassing is a time problem. You can make a core and mold for a large cast-iron casting using a fairly high-volume, gas-producing binder such as pitch (gilsonite) and get away with it.

By the time the pitch is heated to a point where it is producing gas (vaporizing), a protective skin of impervious, solidified metal has already formed. Therefore the gas cannot penetrate through this skin and be absorbed by the molten metal lying directly behind this chilled skin. And, as the gas produced will take the path of least resistance to the atmosphere, it will go out through the mold body and the core vent—causing no harm.

With copper casting where this protective skin is not formed, both the core and mold material must have a minimum of gas-forming ingredients and be permeable enough that the least path of resistance is through the mold body and the core vent. The core and the mold surface can be coated with a good core wash with a low permeability. The skin on the mold surface and the core surface will help because it will offer resistance to core and mold gasses to go in the direction of the molten metal. Of course, any wash chosen must itself be low in gas-producing ingredients. If not you would defeat the purpose. See Fig. 2-4.

Generally, the more readily a material will conduct electricity, the more readily it will conduct heat.

Next to silver, copper is the best conductor of electricity, and next to silver, copper is the next best heat conductor. The use of copper on the bottom of cooking utensils is not for cosmetic reasons. Copper's ability to conduct heat makes for good, even heat over the entire bottom.

As both cores and molds must be quite low in hot strength and permeability, this results in the problem of getting a smooth surface on the casting when pouring a relatively fluid metal against a rather pourous surface. In addition, the hydrostatic pressure exerted by the sprue and or risers combines with the weight of the metal itself, driving it hard against the mold and core contact areas.

It is quite common to use a mold and core wash such as Zirc-O-Graph "A." This is a wash consisting of graphite, zirconium flour with a suitable heat-setting binder. The vehicle is isopropal alcohol. The mold and cores are washed with Zirc-o-Graph by spraying, brushing or dipping. The alcohol is ignited and burns off setting the binder, leaving the mold and or core coated with a fine, smooth refractory coating that, due to its thinness, is quite permeable. The casting will be smooth and the sand will easily peal from them. This is a must with large copper castings in green, skin-dried, dry-sand molds, no-bake, cement sand molds, or what have you. Small castings of light section can be poured successfully in fine, natural-bonded sand with good results if the moisture is kept low and the mold is not rammed too hard (venting the drag and cope well with the vent wire).

MELTING COPPER

Copper can be melted in a rotary, open-flame furnace such as made by U.S. Rotary, a reverberatory furnace such as a Reda, or in the cupola. The products of combustion in contact with the melt plus the problems of controlling the furnace atmosphere will—unless you are a very knowledgable and experienced copper melter—simply drive you nuts and result in gas absorption and heavy oxidation. Your best bet (if you can afford it) would be a high-frequency lift coil induction furnace. The next best bet is a conventional graphite crucible furnace fired with oil or natural gas.

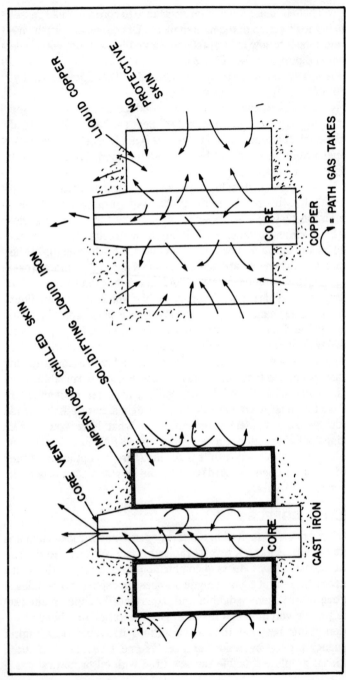

Fig. 2-4. Copper castings do not form a protective skin during solidification. This makes them easily gassed by the mold and core gas.

24

You want to melt as fast as possible, cutting down the time the melt is exposed to the furnace atmosphere. Place the copper in the crucible cold in a cold furnace, cover it with several inches of walnut-size dry, dust-free charcoal. Never put the charcoal on the bottom of the crucible with the metal on top. Note that some foundrymen will take exception to my calling out the use of dry charcoal as a cover for copper when melting. There are two schools on this issue. Some believe that this practice is valueless and, in fact, down right harmful. Melting with a charcoal cover with the burner set slightly oxidizing, the carbon monoxide formed will combine with the water vapor present and yield hydrogen, which will combine with the copper. It is felt by some that no cover should be used whatsoever. A slightly oxidizing flame has a small excess of air over fuel. At this point the feeling is that the carbon monoxide, water vapor, nitrogen and small amounts of oxygen is so insignificant that it is of little or no consequence.

Now you have two choices with or without. Tons of gas-free, high-grade copper has been poured with and without a dry charcoal cover. One shop I worked in (they specialized in medium to very large copper castings) poured every melt of copper through a hot bed of charcoal. See Fig. 2-5. So there you have it. Do what works best for you.

If you are melting with oil, cover the crucible with the bottom of an old crucible or crucible cover. See Fig. 2-6.

Light the furnace (when you have good ignition from the furnace wall) adjust the furnace for maximum firing, and adjust the atmosphere—slightly oxidizing. This is done with an ORSAT, which is a portable apparatus used to determine the amount of carbon dioxide, oxygen, and carbon monoxide in the flue or exhaust gasses from the furnace.

The ORSAT operates on a simple chemical reaction using three reagents. A measured volume of the gas in question is successively passed through three tubes. The first tube contains potassium hydroxide to absorb the carbon dioxide, the second tube contains alkaline pyrogallol to absorb the oxygen, and the third tube contains cuprous chloride in hydrochloric acid to absorb the carbon monoxide. Diminution of the volume after the gas has passed through the three reagents indicates the quantity of each constituent gas, O, CO, and CO_2.

Don't let the word pyrogallol throw you. It is porogallic acid $C_6H_3(OH)_3$, a powerful reducing agent. The alkaline solution

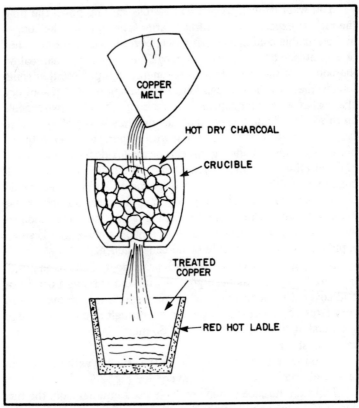

Fig. 2-5. Treating copper by pouring it through hot charcoal.

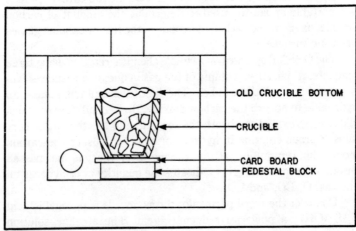

Fig. 2-6. Basic crucible melting arrangement for melting copper enlarged.

26

rapidly absorbs oxygen. Pyrogallol is widely used as an antioxidant in motor oils.

An ORSAT is a handy instrument to own, and it will save you money in fuel bills and enhance casting quality. Without an OR-SAT your best bet is to adjust the furnace (called tuning) to the point where it makes the maximum amount of combustion noise. At this point, you decrease the fuel input or increase the blast air input, either one, only slightly. This will give you a slightly oxidizing atmosphere. This is where you should melt the copper.

Using a decibel instrument to find the maximum level of combustion noise and checking results with an ORSAT, you would be surprised how accurate the tuning process is for maximum combustion. For checking, a cold piece of shiny zinc is held momentarily in the exhaust. If the zinc remains bright you are oxidizing, if it gets smoky the furnace atmosphere is reducing.

Now melt rapidly and do not soak the heat (hold in furnace after it has reached pouring temperature). To avoid soaking, have all your molds ready to pour. A slow-melting crucible furnace or natural-draft, coke-fired crucible furnace should not be used to melt copper. No flux is required with melting copper in the crucible. Simply cover with dry charcoal.

DEOXIDIZING

For deoxidizing, you have basically three choices if the castings are to have maximum electrical conductivity: lithium, calcium boride, and phosphorus.

Let's look at each as a deoxidizer. Copper will, when molten, combine easily with oxygen to form copper oxide CuO, a black insoluble substance formed by heating copper in an excess of air or oxygen. Copper oxide has a melting point of $1235\,°C$, thus copper oxide in the melt. Try as you may you are going to get some.

This oxide of copper must be reduced back to oxygen-free metallic copper. In order to do this, you must introduce something into the copper, carrying these CuO's, that has a much greater affinity with oxygen than the copper. It will rob the CuO particles of the O molecule, leaving the copper behind, and at the same time not contaminate the copper affecting its purity and thus its electrical conductivity.

Lithium is the lightest of all metals. The specific gravity is only .534. It is extremely unstable as a metal when in the presence of oxygen. It will combine with oxygen extremely fast, burning with a dazzling white light in the air when the metal is heated a few degrees above 186 °C or about 335 °F. Below its melting point, it rapidly oxidizes when exposed to the air. It must be kept submerged in kerosene.

For the foundry trade, it is sold in evacuated sealed copper tubes to be used in deoxidizing. See Fig. 2-7.

Now, let's talk about calcium boride as a deoxidizer of copper. Calcium boride is a binary compound of calcium and boron. Boron occurs as borax and boric acid. When combined with metallic calcium, forming calcium boride, it is widely used as a deoxidizer and is extremely useful. It also increases the conductivity of the copper.

Calcium boride is purchased as a black powder from any chemical supply house and some foundry supply houses. Most deoxidizers used with high-conductive copper are sold in pellet form by supply houses. The tubes of pellets are simply calcium boride, and you are going to be told that their deoxidizer is a secret formula. Never will they mention that it is 90 percent to 100 percent calcium boride. The calcium boride is placed in thin-walled copper tubing about 2 inches in diameter and the ends are crimped shut. Each tube contains about 5 ounces of calcium boride powder. See Fig. 2-8.

PHOSPHORUS

White phosphorus (P) actually is yellow in color. Phosphorus ignites spontaneously in air. It unites with oxygen in the air with such rapidity that the chemical action of its combining causes sufficient heat to combust.

Phosphorus has a great affinity with oxygen and therefore is a great deoxidizer for nonferrous metals. It is a very important tool in foundry work because of its property of combining and greatly modifying the characteristics of metals. It is also used in small amounts in steel to increase the strength and resistance to corrosion.

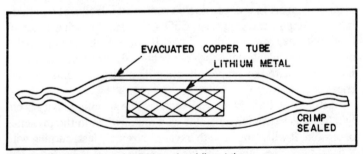

EVACUATED COPPER TUBE
LITHIUM METAL
CRIMP SEALED

Fig. 2-7. Cross section of a lithium deoxidizer tube.

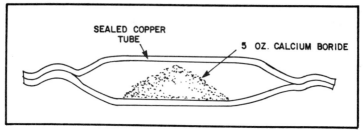

Fig. 2-8. Cross section of a calcium boride deoxidizer tube.

In the old days, we used stick phosphorus to deoxidize red brass heats. It was kept under kerosene and it was extremely difficult to use. You would have to fish out of the kerosene the amount you needed and get it ducked under the melt before it had time to dry out and hook into enough oxygen in the air to cause it to ignite. A phosphorus burn is very bad and it is extremely slow to heal.

Phosphorus is now combined with copper shot and wafer for stability, and it is sold as a deoxidizer for red metals. The most common grades are 10 percent phos. copper—which is 10 percent phos. 90 percent Cu—and 15 percent phos. copper, which is 15 percent phos 85 percent Cu. Its ability to combine with copper to form an alloy of phosphorus and copper is what excludes it as a deoxidizer for pure copper castings.

Even if you could estimate the amount of phos. copper required to deoxidize or combine with the absorbed oxygen in a given copper heat, you would be faced with a residual amount of the phosphorus combining with the copper, altering its characteristics, hardness, electrical conductivity, tensile strength, etc.

Aside from using phos. copper as a deoxidizer, it is used to add phosphorus to bronze to produce phorphorus bronze, defined as any copper-tin alloy with or without lead that has a residual phosphorus content of not less than 0.10 percent.

In view of the fact that phosphorus combines with copper to form the alloy of Cu_3P used as a deoxidizer for high-conductivity, pure-copper castings it is an unwanted impurity.

TEST FOR GASSY COPPER

Now let's settle on calcium boride as our deoxidizer. We must determine first if the copper melt is gassed (and if so how bad). Then we must deoxidize with calcium boride and then determine if we were effective in our treatment.

If you cut several 2-inch diameter blind sprues in a dummy cope and pour these full of the copper from our melt with a small, heated dip-out crucible and observe them as they solidify, this will give you a good indication of the condition of our melt in regard to gas absorption. Also badly oxidized copper is sluggish, dull looking and it has a very poor fluidity.

See Fig. 2-9. Ram up a dummy cope flask (or a box) not too hard, strike off the top, and cut several 2-inch diameter blind sprues about 3 inches deep (test cocks).

Because this first-time test is a familiarization experiment, do not deoxidize the melt. You must be able to recognize the problem when you see it.

With your test cocks prepared, melt your copper as rapidly as possible with the burner adjusted, slightly oxidizing. When you reach a temperature of 2300 °F, test the copper. During the melting, place your test pouring crucible in the furnace exhaust so that it will be hot and not chill the melt when you take your test sample or samples. See Fig. 2-10. Dip out a sample of the melt and pour one or more of your test cocks to within 1 inch of the top.

Watch these. The metal, if badly gassed, will spit violently due to the release of the absorbed gas as the metal cools, and the test cocks will not shrink but form a mushroom-shaped head. See Fig. 2-11.

A sectioned test cock will show the porosity cavitied formed by the gas that could not escape and was trapped by the solidified metal. See Fig. 2-12.

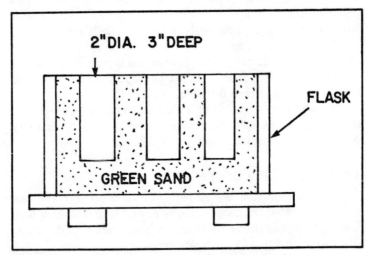

Fig. 2-9. Typical test cock set up to check for gassy metal.

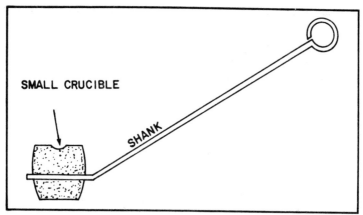

Fig. 2-10. Metal sampling ladle.

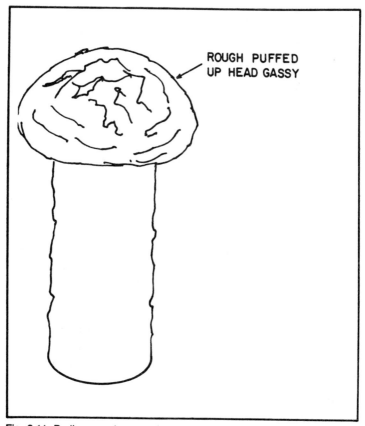

Fig. 2-11. Badly gassed test cock.

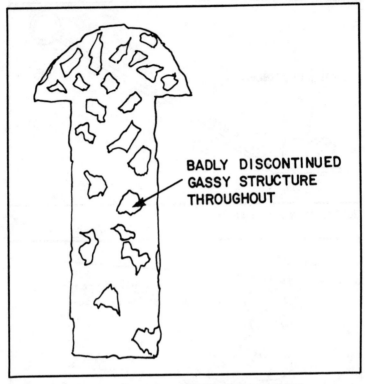

BADLY DISCONTINUED
GASSY STRUCTURE
THROUGHOUT

Fig. 2-12. Section through a test cock showing gassy structure.

If the copper had been deoxidized properly, the test cocks would have shown the characteristic shrinkage and the absence of gas cavities (blow holes). See Fig. 2-13.

It goes without saying that if a mold sprue spits and mushrooms up and does not shrink (as shown in Fig. 2-12), you have poured a bad casting and might as well just toss sprue, gates, and casting back in the pot. Cutting the sprue, casting, or x-raying it is all postmortem; you blew it. This is the reason for testing. There is no excuse for pouring a mold full of gassy oxidized melt. And with copper, you always—regardless of conditions—test prior to pouring molds.

TEST PATTERN

To save time, you can make a test pattern, and make up and keep on hand test molds. These should be core molds, no-bake, or dry-sand molds. My choice is a simple dry-sand core mold using

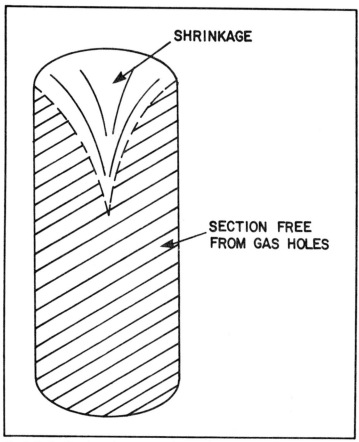

Fig. 2-13. Section through a test cock from a melt that was properly deoxidized.

a simple oil-sand mix and to oven dry them. You need to turn up two dump core boxes (one for the pattern and a bottom core). See Fig. 2-14.

Each mold consists of the two pieces (the bottom core glued to the cavity core). See Fig. 2-15.

TESTING AND DEGASSING PROCEDURE

The copper is melted as fast as possible in a slightly oxidizing furnace atmosphere. When the copper is at 2200°F, one calcium boride deoxidizer tube (5 ounces of calcium boride) is submerged into the melt. The reaction is violent so be sure to allow room for the reaction. If the crucible is too full, you are going to have a mess

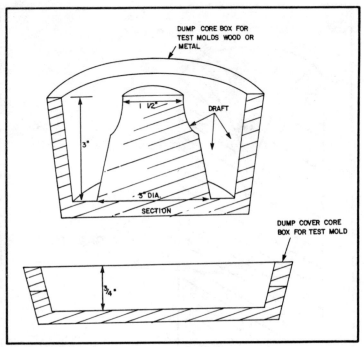

Fig. 2-14. Pattern blue print for dry-sand test mold.

on your hands; 2 inches from the top of the crucible for the metal level is usually sufficient.

When the reaction subsides, mix the metal well with an ordinary iron stirrer. Then skim off the melt. It should, if properly deox-

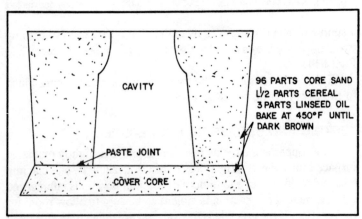

Fig. 2-15. Section through dry-sand test mold.

idized, present a bright mirror appearance. Now, from the treated melt, pour one of your test molds to within 1 inch of the top. If it spits and mushrooms, repeat the calcium boride treatment. When your test casting shrinks and does not spit and gas, pour your castings with this heat.

The tool used to push and hold the calcium boride containers under the melt is a simple iron rod split in a "V" shape on one end. You wedge the tube in place. See Fig. 2-16.

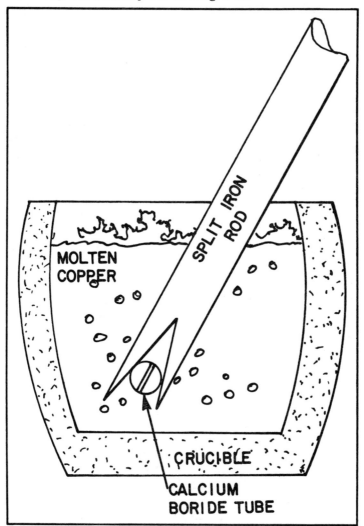

Fig. 2-16. Deoxidizing method with calcium boride.

If you choose, you can tie the calcium boride degassing tube to a rod with cotton twine. Make several because once you use one, it is too hot to handle to try to tie on another tube. The string will not burn if you plunge it below the surface of the melt quickly, excluding the oxygen necessary for the combustion of the string. Don't tie them on with copper wire as the wire melts and the degasser tube will simply pop up to the top.

POURING COPPER

Pour hot 2200 °F for heavy and medium castings and 2350 °F for light, thin work. Pour close to the sprue and pour hard. Avoid gating that nozzles the metal in or a gate that directs the metal against a core. This breaks the stream, increasing the surface area exposed to any oxidizing atmosphere present or generated in the mold cavity.

Your test setup will be used to detect the presence of dissolved gas not only in pure copper but some copper base alloys and aluminum and aluminum alloys. All test castings should, when they have solidified, be cooled at room temperature in water and band sawed for examination. See Fig. 2-17.

Thin copper castings for a large area very often have to be poured two up from both ends. See Fig. 2-18.

A good test casting is not a complete guarantee that the metal from the properly deoxidized melt will result in a gas-free casting. Gas blow holes can be caused by inadequate fillets, wet sand, too low of a permeability (mold or core), wet cores, improperly vented cores, hard ramming, excessive fines in molding sand, pouring practice, pouring too hot, melt contaminated with sulfur dioxide or

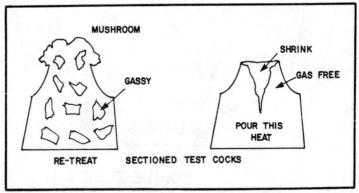

Fig. 2-17. Cross section through two test mold samples: one bad and one good.

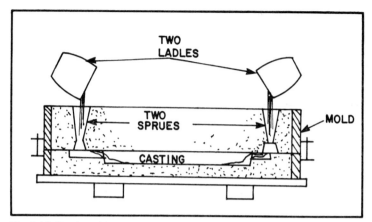

Fig. 2-18. Pouring with two separate ladles in order to run large or thin castings.

hydrogen, metal containing impurities, wet ladles, rigging, wet chill, hot flash, cold sand, hot sand cold flash, and other causes.

Aside from the metal being right, everything else also has to be right.

IMPURITIES

Almost any impurities in a pure copper heat is injurious. Phosphorus, silicon, aluminum, antimony, iron, sulfur, and magnesium are the big offenders.

MACHINABILITY

Pure copper (like most pure metals, aluminum, etc.) are tough materials to machine. This is also a problem when removing gates, risers, flash, and grinding the ingates.

Grinding wheels are offered to the trade for grinding nonferrous metals. These wheels are especially formulated to break down fast, run cool, and not glaze or load up.

If a wheel starts to load up and glaze at normal use, you can sometimes correct this by increasing the speed of the wheel or use a softer wheel. A grease stick (tallow) touched against the wheel for lubrication also helps. Sawing copper is also a difficult proposition. You need blades with not over 14 teeth to the inch for hand sawing with at least a 10-degree tooth rake. For power hacksaws, 10 to 14 teeth with a 10-degree rake will do the trick. See Fig. 2-19.

Milling saws should also have a 10-degree tooth rake and the sides of the teeth should be relieved for clearance. A pitch of a 1

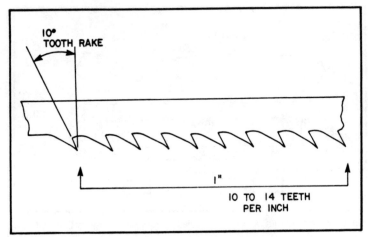

Fig. 2-19. Correct hacksaw blade for sawing pure copper.

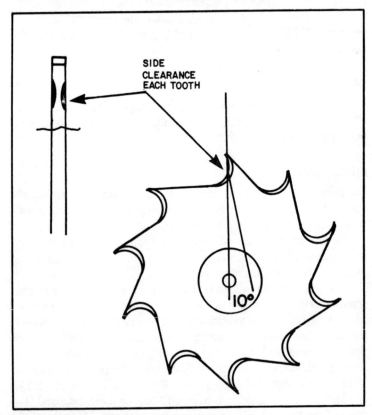

Fig. 2-20. Correct milling saw for pure copper work.

38

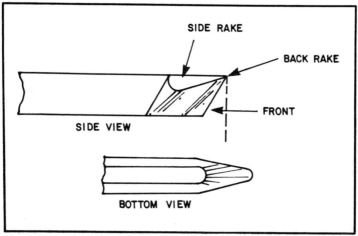

Fig. 2-21. The correct sharpening for a lathe tool suitable for casting pure copper.

inch would give you 10 teeth on a 3-inch diameter milling saw. See Fig. 2-20.

Lathe tools for turning pure copper should have a front clearance of 7 degrees a side clearance of 7 degrees, a back rake of 10 degrees a side rake of 25 degrees, and turning speed 75 to 150 feet pr minute. See Fig. 2-21.

Side clearance is to permit the cutting edge to advance freely without the heel of the tool rubbing against the work. The front clearance is required to let the tool cut freely as the tool is fed into the work. It stands to reason that if this angle is too great, the cutting edge will break due to insufficient support. You must find the correct angle for the material you are machining. See Fig. 2-22.

Side and back rake is also to facilitate free cutting for hard

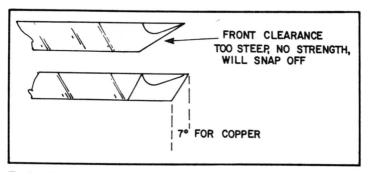

Fig. 2-22. Proper tool cutting edge, and clearance for lathe turning pure copper.

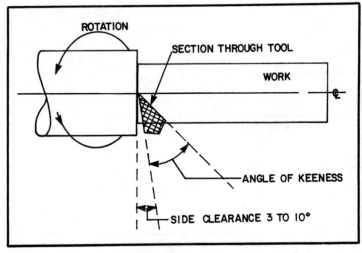

Fig. 2-23. Lathes angle of keenness (illustrated by cross section).

bronze, cast iron, and steel. You need very little side rake or back rake. More is required for softer material.

A point little talked about is the angle called the angle of keenness. This angle can vary from as much as 60 degrees for soft material to nearly 90 degrees for hard bronze, cast iron, and steel. See Fig. 2-23.

The cutting edge should be exactly on center when machining brass, copper, aluminum and any other tenacious metals. See Fig. 2-24. For most other turnings, the cutting edge is set 5 degrees above center. See Fig. 2-25.

The cutting edge of a tool should be exactly on the center when boring, taper turning, and when cutting screw threads (regardless of the material being machined).

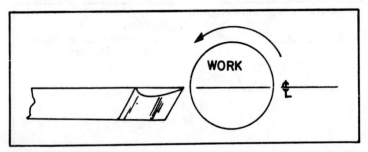

Fig. 2-24. Tools relationship to the center line of the work being turned, when turning pure copper.

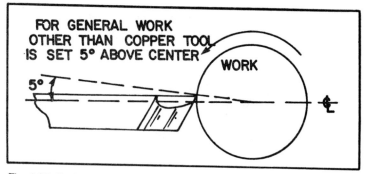

Fig. 2-25. Tool setting for normal turning other than pure copper.

Cutting tools for copper should be extremely sharp because any saw tooth edge (or rough edged) tool will drag, dig in, and gouge. This goes for drills, taps, boring bars, and what have you. If you are not into machining soft copper, it would be wise to look up some old machine shop operator and get him to lay the word on you. Getting information out of your tool supplier is not what is used to be. These fellows today are for the most part only sales people. They have very little or no industry knowledge. Beware of the high-technology people.

GATES AND RISERS FOR COPPER

Copper has a high shrinkage: 1/4 inch per foot is required for the patterns for medium and heavy work and 3/16 of an inch is need-

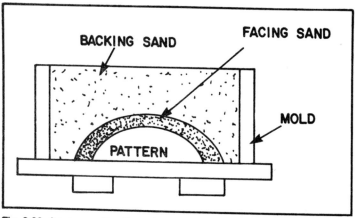

Fig. 2-26. A cross section through a sand mold showing the use of a facing sand against the pattern.

ed for light work. You have to ensure adequate feed metal so risering is considerably larger for copper castings.

FACING

Usually, no facing sand is required for light work because the sands used are fine enough to give a nice finish. Nevertheless, as you get into heavy and medium work where you require a courser sand to give you the necessary added permeability, you can use a finer sand as a facing and back with the courser sand to give you the finish you desire. Of course, if your facing layer is too thick you will defeat the purpose of using a high-perm sand for heavy work. You should never need a facing of more than an inch thick at the most. Try for 1/4 of an inch. See Fig. 2-26.

Pure copper castings for electrical work and various uses where pure copper is required is a speciality game. One who specializes in this class of work, even on a small scale, will never be without work. It is, however, a tough nut from every aspect—casting and machining.

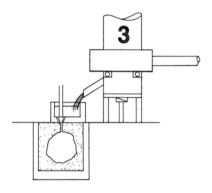

Red Brass

The term red brass can be very confusing because the standard wrought metal called red brass is actually a rich, low brass composed of 85 percent copper and 15 percent zinc. This is the brass used for pipe, stampings, and drawing. Annealed it has a tensile strength of 35,000 PSI and an elongation in 2 inches of 40 percent. More than one fledgling founder has purchased this material from the junkyard only to find that when he melted it, a large percent of the zinc went off as zinc oxide and the shop looked like it was snowing with all the ZnO floating around. Don't confuse this material with red brass used for casting.

Leaded red brass and leaded semired brasses are reasonably priced and are easily cast and machined. They have relatively good pressure tightness, they polish to a fine finish, they drill and tap like butter and they have good strength. Their application or uses cover thousands of items. Plus red brasses are reasonable in cost.

The only true red brass (casting brass) is the alloy of 85 percent Cu, 5 percent Sn, 5 percent Pb, and 5 percent Sn, called ounce metal. Commercial red brass, leaded semired brass, 78 semired brass, and 76 semired brass are all commonly called red brass. It seems like anything that looks like red brass is called red brass.

The foundry practices for all five of these alloys is very much the same with little variation. It would be repetitious to cover each separately. So let's look first at the nominal chemical compositions and properties of each separately.

As you can see in Table 3-1, there is little difference between 85,5,5,5 and 83,4,6,7. This brass is the same hardness at the 85,5,5,5 but a little weaker in strength. Neverthless, it is easier to machine. The figure I am using for machinability is based on 100 (machines like butter), 90 (almost like butter) down to a copper nickel alloy of 90 percent Cu 10 percent Ni (which is rated at 10 for machinability, tough bad-news metal to machine). With pure copper, I would give about an 8 on the scale of 1 to 100.

The 78 semired brass and the 76 semired brass (Table 3-2) are also called plumbers special or PS brass used to cast light, simple plumbing ware. The "P" traps and "T" parts are for toilet valve mechanisms, novelty items, and anything that calls for simply an inexpensive brass that is easy to cast, machines easily, polishes easily, and plates easily. General odds and ends (such as coat hooks) work.

FOUNDRY PRACTICE

The 85-5-5-5, 83-4-6-7, 81-3-7-9, 78-3-7-12, and 76-3-6-15 are

Table 3-1. Red Brass Properties.

Name & Composition	Properties	
Red Brass 85% Cu 5% Sn 5% Pb 5% Zn	Weight, lbs/in^3 Patternmaker's shrinkage Solidification range Pouring temp. light work Pouring temp heavy work Tensile strength Yield strength Elongation % in 2" Brinell hardness (500 kg) Machinability Heat treatment	0.318 3/16" per ft. 1850-1570° F 2100-2350° F 1950-2150° F 37,000 PSI 17,000 PSI 30% 60 84 No response
Commercial Red 83% Cu 4% Sn 6% Pb 7% Zn	Weight lbs/in^3 Patternmaker's shrinkage Solidification range Pouring temp. light work Pouring temp heavy work Tensile strength Yield strength Elongation % in 2" Brinell hardness (500 kg) Machinability Heat treatment	0.312 3/16" per ft. 1840-1550° F 2100-2300° F 1950-2150° F 35,000 PSI 16,000 PSI 25% 60 90 No response

Table 3-2. Red Metal Mixes.

Oil Sand Mix	
Washed & dried silica sand	95.5 lbs.
Corn flour	.2 lbs.
Western Bentonite	3.0 lbs.
Dextrine	1.3 lbs.

Oil Binder Core Mix	
Washed & dried silica sand	97.0 lbs.
Corn flour	0.5 lbs.
Kerosene	1.0 lbs.
Linseed Oil	1.5 lbs.

five red brasses composed of the same copper, tin, lead, and zinc. Different percents account for the variation in the properties.

The lower the copper and tin content the less the metal will cost you as ingots or as scrap. When buying from your local friendly junkyard, you will be charged for 85,5,5,5 regardless of what you receive. And if the yard buys from you, they will pay you for semired. Unless you know what you are buying and stand your ground, the yard owner will get to you. It's hard to tell one grade from another by looking at it.

When buying scrap, the item itself will sometimes provide a clue to its general composition. Steam valves are most likely 85 3-5's; nameplates, builders' cast hardware, and door knobs are most probably cheap red or semired. If you purchase considerable brass scrap, it might pay you to set up enough simple lab equipment to determine the copper content of the material you want to buy.

Knowing the copper percent is a very good clue as to what the alloy should be. Suppose you have a chance to buy a 500-pound lot of globe valves, and the color is that of red brass. Are they high grade ounce metal, PS, or what? Without going through a complete chemical analysis (quantitative) you could simply determine the copper content of a red-metal alloy.

COPPER CONTENT

Weigh out 0.25 grams of drillings into a 300cc flask and dissolve these drillings in a small quantity of dilute nitric acid HNO_3. A half-and-half solution is good half-nitric acid by volume to half distilled water by volume. **Caution**: add the acid to the water; never add water to an acid.

You can warm the solution slightly. This will help to get your drillings into solution. When the drillings are into solution, the solution will be blue to blue/green in color (as copper ions are colored). Now carefully boil the solution down to expel an excess of nitric acid, take the flask off of the heat source, and slowly add pinch-at-a-time sodium carbonate until the solution is neutralized (no longer acid). You check with litmus paper (red is acid and blue is basic).

To the neutralized solution, add drops of acetic acid until the solution is slightly acid (red on litmus) and allow the solution to cool. Add potassium iodide to the solution and titrate with 0.1 m sodium thiosulphate using starch as an indicator: 1 cc/0.1m sodium thiosulphate equals 0.0063 gram copper.

SOLUTIONS

Sodium thiosulphate 0.1m can be purchased or made by dissolving 24.8 grams of sodium thiosulphate ($Na_2S_2O_3.5H_2O$) in distilled water and dilute to one liter.

Potassium Iodide solution can be purchased or prepared. To prepare potassium iodide solution, dissolve 16.6 grams of potassium iodide (KI) in distilled water and dilute to one liter. This gives you a 0.1m solution.

For your starch solution (starch indicator), make a thin paste with one gram of starch. Pour this into 100 milliliters of boiling water while stirring.

With this simple setup you can determine the copper content. With this knowledge, coupled with a visual and physical examination of the material, you can come very close to knowing what you have. If the junkman says its 85,5,5,5 and your simple chemistry test shows by percent weight that the copper content is, for example, between 78 percent to 76 percent you have been sold semired brass. Further down the line, say 72 percent to 63 percent, you are into yellow brass.

The description just given is simply to illustrate that you should have some chemical understanding in regard to simple quantitative analysis of the metals with which you are dealing. Do not let chemistry intimidate you. It is quite simple when you understand the basics.

FOUNDRY MANIPULATION

The five leaded red brasses and semileaded red brasses are very commonly used casting brasses. The list of items cast in these five alloys would be endless. To carry all five of these alloys in the foundry would not be very practical and would tie up money for no good purpose.

With commercial red 83-4-6-7 and semired 76-3-6-25, you have covered 99.999 percent of the work covered or suitable for red brass. The 83-4-6-7 is for valves and pressure work and general machinery castings, and 76-3-6-15 is for novelties and small parts.

There is quite a price spread between 85-5-5-5 and 83-4-6-7, but their typical properties are very close. The price spread between 85-5-5-5 and 76-3-6-15 is quite wide.

The 85-5-5-5 is probably the oldest and most widely used alloy in the red and semired group. When someone mentions or calls for red brass as the required alloy for a particular casting or castings, you can bet 85-5-5-5 is what they require. It goes without saying, when the spec. simply calls for red brass, that it is 85-5-5-5.

In most cases the customer or designer will call for ounce metal regardless of the intended use or physical properties required by the casting. I have seen 85,5,5,5 specified as the alloy required for items such as a coat hook, door knocker, or perhaps a simple plaque. This is where the foundryman gets into the act. He should advise the customer about which alloys are less expensive but still suitable.

MOLDING SAND

Any good, fine natural-bonded molding sand or synthetic molding sand would meet the requirements for the weight and section thickness of the casting. Castings are from a few ounces up to 1 pound, not over 5/8 of an inch in section thickness. Permeability is 20, green compression is 7 PSI, and moisture percent is 6 to 6 1/2 percent.

Castings are from 1 pound up to about 15 pounds, with a section thickness of 1 inch or less. Permeability is 30/35, green compression is 7 PSI, and moisture percent is 5.5 percent to 6 percent.

From the point up to 50 pounds with a section not exceeding 2 inches thick, permeability is 40, green compression is 7 PSI, and moisture percent is 6 percent Max(maximum).

Although red brass castings of 1000 pounds and upwards have been successfully cast in green sand, if you go over 50 pounds in weight and 2 inches in section, you are much safer to make the casting in dry-sand molds, core molds, or one of the no-bake systems available such as furan, CO_2, etc.

If you lose a small casting or even a bucket full, it's really no problem. When you lose a really big one it is a very costly problem. In some cases, just trying to cut it up into bite-size pieces so you can remelt is next to impossible. At any rate, when you get up there in size in green sand you are looking at green strengths of 15 to 20 PSI and permeability in excess of 95 with moisture 5 percent maximum.

When you make a green-sand mold, the heavier the casting the larger the problems. With a dry mold (no-bake, etc.), you can achieve lots of strength, high permeability, near zero percent moisture, along with good finish, minimum or no burn in, good mold collapsability, and at the same time good hot strength.

FACING

You should not require any facing for the average red-brass castings; they should be smooth and peel nicely if your green sand is kept in shape and is refractory enough. In special cases such as plaques, grave markers and highly ornate work, you might want to use a fine facing. You might dust the molds with wheat flour (in an old sock or parting bag) for light to very light work. Some molders dust all molds with flour. It is beneficial and the cost and time is small.

Finely powdered rosin can be dusted on the mold cavities. When the incoming hot metal comes in contact with the rosin, it decomposes and throws off a dense black smoke. This smoke coats the sand surface, and gives you the same results as if you were to smoke the mold with acetylene. This usually results in a smooth casting of good color.

On fine high-detailed work such as antique reproductions, the practice of printing back is used. The pattern is drawn and the mold surface is dusted with graphite or portland cement. Then a dusting with finely powdered charcoal, the pattern is returned and bumped down to set the facing of graphite or cement firmly and smoothly against the sand. Then the pattern is removed again, and the mold is closed and poured. See Fig. 3-1.

The purpose of the coating of charcoal on top of the graphite or portland cement is to keep it from sticking to the pattern when you return the pattern to the cavity (print back). If you fail to do this, you will lift some or all of the graphite or cement off of the surface when you make the final draw of the pattern (defeating what you are trying to do).

My preference is to use portland cement. This is done where

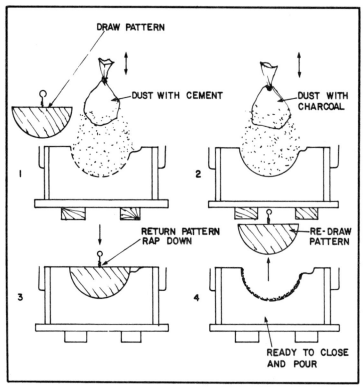

DRAW PATTERN

DUST WITH CEMENT

DUST WITH CHARCOAL

1

2

RETURN PATTERN
RAP DOWN

RE-DRAW
PATTERN

3

4

READY TO CLOSE
AND POUR

Fig. 3-1. Step by step illustration of printing back a pattern to produce a very smooth sharp surface to the mold and casting.

you have mostly horizontal surfaces. You can't get it to work very well on vertical areas. If done properly it is amazing as to the detail and sharpness you can achieve (great for plaques and medallions).

This process can be used with all red metals as well as cast iron. This is the way highly detailed work can be cast on woodstove parts and trim. Now and then you see an old, very ornate cash register with iron or brass castings covered with floral designs.

GATES AND RISERS

With most light red brass castings that are light in weight, (a pound or less) and the section is uniform in thickness throughout, all you need is a simple choke gate. See Fig. 3-2.

Of course, larger castings might require risers and chills. In many cases I have seen that castings were defective not from too simple or small gates but from overgating too large and too many.

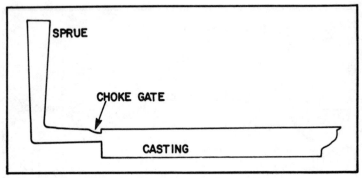

Fig. 3-2. Cross section through a simple choke gate.

In many cases, the gates were sound land the casting defect was caused by the casting feeding the gate or gating system rather than the other way around.

The tendency for the novice is to overgate as insurance against losing the casting to misruns or what have you. In many cases, a molder will enlarge the gates to correct a misrun problem when the fault is caused by cold pouring, improper pouring, bad design, pattern equipment, hard ramming,low permeability, excessive moisture, etc. See Chapter 12.

CORE PRACTICE

There are lots of formulas and various binders available. A good no-bake (furan) core will do very well in most cases. The old standby for most red metals is the simple oil-sand mixes with 1 part of oil to 30 parts of sand (a good washed-and-dried, sharp-silica sand). If you need some green strength, you can add some wheat flour, cereal flour, or you can simply add 50 percent bank sand that has a clay content as high as 6 percent to your clay-free core sand. Another alternative is to add a small amount of southern bentonite to your mix. Clay in a core sand mix will absorb oil readily, and you must account for this by raising your oil ratio accordingly.

For 90 percent of your core work with red brass (85,5,5,5) manganese bronze, tin bronze and leaded tin bronzes, Table 3-3 mixes will produce quality work.

The sand in either case should be 60 to 80 AFS grain fineness with less than 1 percent clay content. A good grade of commercial core oil can be substituted for the linseed oil. The moisture should be no higher than 3 percent.

The cores are baked at 350 °F to 400 °F until dry (a light to medium brown color). The two core mixes given will also serve equally as well for complete dry-sand molds. Cores for red metals usually do not require a core wash except for large castings of heavy section or the area where the core is quite thin, such as a bib valve or small globe valve where the core goes through the seat and is strengthened at this point with a wire. The core need only be washed at this point with a commercial wash or a wash made of plumbago with a 10 to 1 molases water, 10 parts water, 1 part molasses. See Fig. 3-3.

There are two reasons for a wash here. A small amount of core material at this point is subjected to considerable heat that could

Table 3-3. Semi-Red Brass Properties.

Name & Composition	Properties	
Leaded Semi-red Brass 81% Cu 3% Sn 7% Pb 9% Zn	Weight lbs/in^3 Patternmaker's shrinkage Solidification range Pouring temp. light work Pouring temp. heavy work Tensile strength Yield strength Elongation % in 2" Brinell hardness (500 kg) Machinability Heat treatment	0.314 3/16" per ft. 1840-1540° F 2100-2300° F 1950-2150° F 34,000 PSI 15,000 PSI 26% 55 90 No response
78 Semi-red Brass 70% Cu 3% Sn 7% Pb 12% Zn	Weight lbs/in^3 Patternmaker's shrinkage Solidification range Pouring temp. light work Pouring temp. heavy work Tensile strength Yield strength Elongation % in 2" Brinell hardness (500 kg) Machinability Heat treatment	0.312 3/16" per ft. 1790-1540° F 2100-2300° F 1950-2150° F 35,000 PSI 14,000 PSI 28% 55 90 No response
76 Semi-red Brass 76% Cu 3% Sn 6% Pb 15% Zn	Weight lbs/in^3 Patternmaker's shrinkage Solidification range Pouring temp. light work Pouring temp. heavy work Tensile strength Yield strength Elongation % in 2" Brinell hardness (500 kg) Machineability Heat treatment	0.310 3/16" per ft. 1750-1530° F 2100-2300° F 1950-2150° F 36,000 PSI 14,000 PSI 30% 55 90 No response

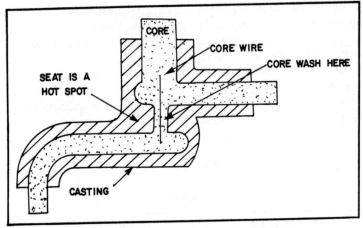

Fig. 3-3. Cross section through a bib casting and internal core.

cause it to break down (collapse prematurely). Also, if the core wire is exposed at any point (which it usually is), the brass will weld itself to the wire if the wire if not coated with a refractory wash.

MELTING

Leaded red and leaded semired brasses should be melted as rapidly as possible and only raised about 100° above the pouring temperature of the job at hand. Note that if you should have a variety of casting weights to pour from the same heat of metal, be sure you mark the molds so that you know what is what. Pour the light work first, medium next, and heavy last.

If the pouring temperature for a given mold is crucial, check the ladle or crucible at the mold with a pyrometer. Should the metal be too hot for the job, don't try to cool it down by adding solidified metal, sprues, etc.

This practice usually results in big problems, gassy melt, cooled down too far, semiliquid chunks in the melt, or—if wet—the addition can cause an explosion. Simply let it cool down of its own accord. When the temperature is correct for the job, skim again and pour.

When melting in a solid-fuel (natural or forced draft) furnace, melt with coke not coal. A good grade of cupola coke should be used. You do not see many pit solid-fuel furnaces anymore. Nevertheless, you should not overlook the pit natural-draft, solid-fuel furnace for the small foundry. It will do a good job. It is inexpensive to build and inexpensive to operate. See Fig. 3-4.

When melting in an oil- or gas-fired crucible (stationary, pit, or tilting) melt the metal under a slightly oxidizing atmosphere; melt as fast as possible. Slow melting can result in gas absorption and high loss due to oxidation.

It is simple logic. The longer you expose the melt to the products of combustion the greater the chance for oxidation and gas absorption. The basic reason for melting with a slightly oxidizing furnace atmosphere is to prevent the absorption of hydrogen in the melt that, when combined with oxygen, forms H_2O (water). H_2O + heat = steam = porosity in the casting. Absorbed hydrogen is extremely difficult to remove from the melt. Oxides that form metallic oxides are easily removed by simple deoxidizing treatment.

So we choose the lesser of the two evils, and one we can easily cope with. Regardless of the type of furnace you melt with, the following general rules apply.

☐ Melt fast.

☐ Under an oxidizing atmosphere.

☐ Super heat only about 100 °F above the pouring temperature of the job at hand.

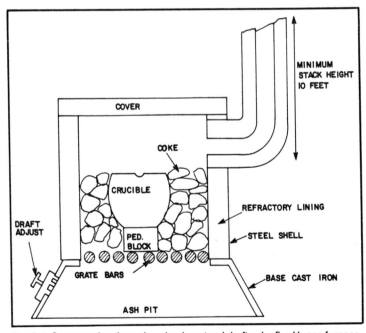

Fig. 3-4. Cross section through a simple natural draft coke-fired brass furnace.

☐ Don't soak melt by holding the melt in the furnace after it is up to temperature. When it's ready, pull it and pour.

☐ Use a pyrometer to check melt temperature. No way can you gauge the temperature by eyeing it.

Melting in any open flame (noncrucible furnace) such as rotary, reverberatory, or indirect arc requires a slag cover to prevent high oxidation of the melt and gas absorption. At any time the bath is in direct contact with the products of combustion, the problems are increased a thousand fold. A reverberatory furnace is a very tricky furnace in which to melt. The bath must be quite shallow, with a large surface area, because it is heated from only one side by the radiation of the heat above it. If the bath is too deep and the slag cover too thick, you will never get it melted or up to temperature. If the bath is too shallow and the slag cover too thin, oxidation and gas absorption will be excessive. In addition, breaks (open areas) in the slag cover will result in huge oxidation losses and a melt that is impossible to completely deoxidize. See Fig. 3-5.

In the rotary (open flame furnace) or the indirect arc (rocking), you are much better off as the heat is delivered to the melt by both radiation and conduction. This allows you to melt a rather deep metal bath under a thick slag cover. See Fig. 3-6.

LET'S TALK ABOUT IT

The rotary furnace is a simply horizontal barrel type of fur-

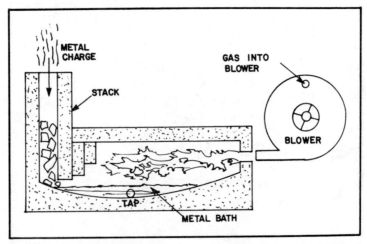

Fig. 3-5. A cross section front view of the basic design of a gas-fired reverberatory furnace.

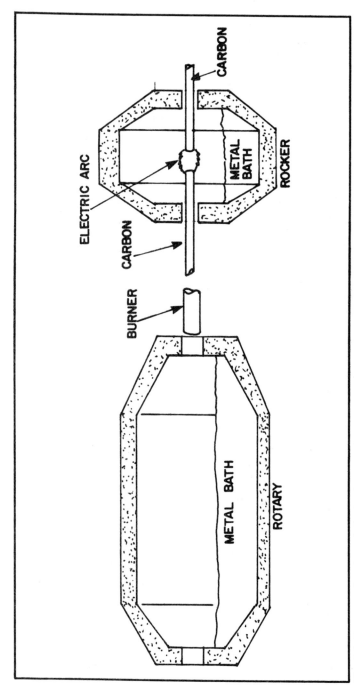

Fig. 3-6. Cross section through the basic rotary furnace and the arc rocking furnace.

nace lined with refractory and fired through one end with gas, oil, or pulverized coal. The furnace construction varies—from a gimbel arrangement, whereby the furnace can be tilted to pour, to a rigid construction—and is tapped similar to a cupola. The big advantage to the rotary or rocker (arc) is the heating of the metal by both radiation and convection. See Fig. 3-7.

As shown in A of Fig. 3-7, the radiant heat is directed down, supplying heat. The lining area above the melt is superheated and is slowly revolving until it reaches the point as shown in B of Fig. 3-7. You have the lining giving up heat by convection to the melt. This heat exchange by radiation and convection is continuous due to the slow rotation.

This combination of radiant and convection heat exchange is a much more efficient system than by radiation alone. Rotaries are available in sizes from as small as 150 pounds brass capacity to 3000 pounds or greater.

A small furnace of 150 pounds capacity will melt (to pouring temperature) 150 pounds of brass in one-half hour; a 500-pound rotary will take only three-fourths of an hour. The 150-pound unit will take about 3 gallons of #2 fuel oil to do the job. The 500-pound furnace will take about 6 1/2 gallons to melt the 500 pounds. With natural gas at 1100 Btu per cubic feet on the small furnace, 150 pounds can be melted with about 200 to 250 cubic feet of gas. For 500 pounds, it takes about 500 cubic feet of gas.

While it takes some doing, a small homemade rotary furnace is a sure winner for the one- or two-man shop. One disadvantage,

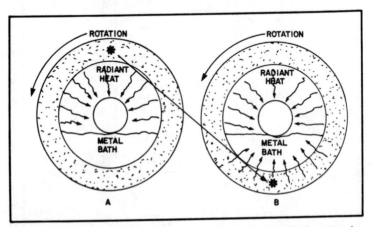

Fig. 3-7. With a rotary or rocker you get both radiant and convection heat transfer to the metal bath.

if you could call it a disadvantage, is that you must transfer the melt to a preheated ladle or crucible for pouring.

CUPOLA MELTING

Leaded red brass, and semired brass can be melted in the cupola, but your lead loss will run from 15 to 25 percent, and your zinc loss could (and will) run from 60 percent upwards. Melting is inexpensive because the coke-to-metal ratio is only 1 to 20 (1 pound of coke will melt 20 pounds of red brass).

An average production for eight hours is about 4500 pounds per hour with a 26-inch-inside-diameter cupola. Of course, the zinc and lead must be replaced. The usual procedure is to melt the copper and teem it into a very hot ladle containing the preheated tin, lead, and zinc.

Alloying in the Ladle. The tin loss in the cupola—should the tin be already alloyed with the copper—is negligible, and the copper loss in cupola melting is zero.

You can build a small cupola or line down one to 18 inches, but here again—unless you need a continuous flow of metal to maintain a highly mechanized molding production or have one huge casting to produce—you are better off with a crucible rig or rotary furnace. A cupola smaller than 18 inches inside diameter is difficult to operate.

The only reason I mention the cupola is that you never know who might want to melt using the cupola. I know of one small operation, a two-man shop, that specializes pouring only life-size and larger-than-life-size bronze art work. They cast only in silicon bronze (which can be melted without loss in the cupola). The mold is prepared in front of the cupola in a pit and poured directly from the cupola. See Fig. 3-8.

One last word on the cupola. The melting of red metals in the cupola is rapid (very rapid), economical, and you can melt just about anything such as borings, turnings, and skimmings. It is easy to obtain hot metal without noise and loss of heat to the surrounding area. The product is clean and high in quality.

FLUXING AND DEOXIDIZING

First let's clear the air on exactly what is meant by *flux* and *slag*. Flux comes from the root meaning of flow. This is the common definition used in the foundry. It is also defined as conversion and motion plus other uses.

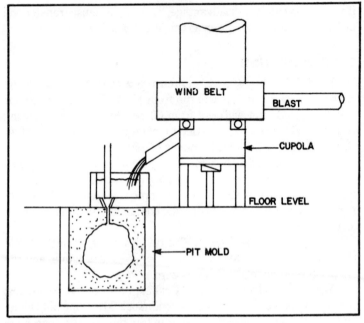

Fig. 3-8. A large mold being poured directly from the cupola.

The woods are full of various recipes for fluxes with claims as to what they will or will not accomplish. There are basically four fluxes: oxidizing fluxes, reducing fluxes, cover fluxes, and tougheners. Let's look at them separately.

An oxidizing flux would be any flux that will introduce oxygen into the melt for the purpose of removing oxidizable impurities, leaving behind the unoxidizable material you want to purify.

A simple case would be where you had a heat of gold that is not pure or not as pure as you prefer. By adding (or using) an oxidizing flux, the metallic impurities present such as copper and lead would become oxidized by the oxygen-producing flux.

These oxides of copper would float to the top and collect in the slag. Then they can be removed by separating the liquid metal from the slag. The most common of the oxidizing fluxes are fluxes where the main constituency is potassium nitrate (saltpeter) and sodium nitrate (chile saltpeter). These nitrates are widely used to supply oxygen necessary for combustion in many explosives such as gunpowder (black powder), which is a simple mixture of 75 parts of potassium nitrate, 15 parts of charcoal, and 10 parts of sulphur.

The potassium nitrate simply supplies the oxygen to support

the combustion of the combustibles in the mix. As an oxidizing flux, it supplies the oxygen to combine with the element you want to flux off as an oxide.

Reducing Fluxes. Reducing fluxes bring about reduction (the opposite of oxidation). Reduction is accomplished in three ways:

☐ By the acceptance of one or more electrons by an atom or ion.

☐ The removal of oxygen (deoxidize).

☐ By the addition of hydrogen.

Reduction or reducing fluxes consist of (or various mixes of) charcoal, sugar, sodium carbonate (soda ash), argol (wine lees, cream of tartar), potassium carbonate (pearl ash), potassium cyanide, wheat flour, etc.

Reducing and oxidizing fluxes are primarily used in metal-refining operations to adjust the composition, by removing and or adding various components. A reducing flux can be of some benefit in melting red metals, but in most cases the benefits are too negligible to be of concern. And, of course, an oxidizing flux is just what you don't want with red metals.

Cover Fluxes. These fluxes are simply covers and neither add oxygen nor take it away. The purpose of cover fluxes is to cover the melt and, in doing so, exclude the products of combustion present from coming into contact with the melt and combining with the melt chemically to form oxides, absorbed hydrogen, etc. The cover fluxes that are borates (borax and boric acids) will dissolve and flux off dirt and sand present. They will, however, also flux off any metallic oxides present. The result is melt losses of copper. The most common of cover fluxes are borax, boric acid, powdered or broken glass, and fluorspar (alone or in various combination and mixes).

Fluorspar (fluorite). It is a crystalline or massive granular mineral CaF_2 (calcium fluoride). Fluorspar is a super flux. It will make an extremely fluid slag and will, if used in excess, flux the lining completely out of a furnace in nothing flat. It can also ruin a crucible in nothing flat. The problem with fluorspar is that it is used to produce hydrofluoric acid (hydrogen fluoride in aqueous solution). The hydrogen fluoride gas from a mixture of calcium fluoride (fluorspar) and sulfuric acid H_2SO_4.

The products of combustion and the atmosphere in the furnace often have the ingredients and the right conditions to produce hydrofluoric acid when fluorspar is present as a flux. Hydrogen, sulfur, oxygen plus calcium fluoride can very easily add up to hydrogen fluoride + H_2O = hydrofluoric acid. This acid will dissolve glass and silica very readily.

It is only used in very small amounts where the material being melted is extremely dirty and sandy, and you have a lot of silica material to reduce to a very liquid slag. If you have a slag or flux cover in an open-flame furnace or the crucible is too thick and gummy to effect a good separation (skim) from the metal—be careful and use only enough to thin the slag out and no more—you can make a very useful tool of fluorspar. You must always use fluorspar carefully.

Tougheners. Ammonium chloride, mercuric chloride, and copper chloride are the so-called toughening fluxes. They are used primarily in purifying gold that is almost fine.

Warning. Ammonium chloride NH_4CL is sal ammoniac and is toxic by inhalation. Tolerance (fume) is 10 mg. per cubic meter of air. Mercuric chloride $HgCl_2$ is highly toxic by ingestion: inhalation and skin absorption (could be fatal!). The tolerance is only 0.05 mg. per cubic meter of air. Copper chloride $CuCl_2$ or $CuCl_2 \bullet 2H_2O$ is only moderately toxic.

Flux chemistry is quite complicated but is of little concern to the small, nonferrous founder. Neverthless, it pays to at least have some knowledge of what fluxes do and how.

For melting leaded red and semired brasses, fluxes are not particularly necessary especially when melting clean materials. That is in a crucible furnace. If the material is dirty, a cover flux of borax and glass (lead free glass) works out fine. The consistency of the flux can be controlled by the borax content.

A much-used combination refining and cover flux is soda ash 5 parts, calcined borax 3 parts, and ground flint glass 1 to 12 parts. Some prefer a 50/50 mix of soda ash and pearl ash instead of soda ash alone.

When melting material in an open-flame furnace such as a rotary, reverberatory, or rocking arc, you have no choice but to cover the melt with a flux to protect it from the products of combustion. The most common flux used is ground glass (flint) and Rasorite.

Rasorite is a brand name for sodium borate concentrates. Rasorite is also sometimes called Kernite. The composition of Rasorite is $Na_2O \bullet 2B_2O_3 \bullet 4H_2O$ and kernite is $Na_2B_4O_7 \bullet 4H_2O$. Both are natural sodium borates mined in Kern County,

California. Rasorite is sold by the U.S. Borax & Chemical Co., 3074 Wilshire Blvd., Los Angeles, CA 90010.

I could go on and on about fluxes, basic fluxes, acid fluxes, neutral fluxes, and their reactions. What works for what?

The big problem with the use of fluxes in crucible melting is that 99.9 percent of the time no flux is required. It seems almost impossible to convince nonferrous melters of this fact. You are melting metal; you are not engaged in the smelting and refining process.

The foundry supply companies actually have hundreds of various fluxes for sale for every purpose in the world (none of which you need). If a melter insists that a flux is needed, they are operating on the theory that if a little is good a lot is much better. No flux available will offset poor melting practice.

If you need a flux because the components of the charge are excessively dirty, if you clean the material before charging, then you don't need a flux.

Crucibles are costly, and the life of a crucible can be greatly reduced by the use of a flux or the injudicious use of a flux. If a flux is required, use only a minimum amount. Do not add it to the bottom of the crucible. Add it to the metal when it is just melted. I have seen new crucibles ruined on the first melt due to the flux action rapidly fluxing away the crucible at the metal level. See Fig. 3-9.

If the furnace is adjusted improperly, excessive oxygen (very oxidizing), aside from the metal loss due to metallic oxides, will

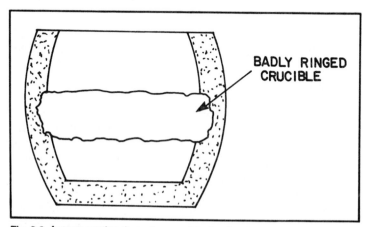

BADLY RINGED CRUCIBLE

Fig. 3-9. A cross section through a crucible that has been ringed badly by the action of a caustic flux.

contribute to crucible attack. This defect is called *ringing* regardless of the cause.

Slag is the product of the fluxing of sand, dirt, metallic oxides, oxides of silicon, the refractory lining, and the crucible. All these and more combine chemically to form a viscous, gummy slag. Slags can be basic, acid, or neutral.

Soda slags are very reactive with ordinary clays and airsetting plastic cements. Although soda slags are highly basic, it is reactive with basic refractories such as chrome and magnesia brick. To minimize ringing of crucibles, use very little flux (preferably none).

With open-flame melting (noncrucible furnace) where you must have a slag cover, you have the problem of the slag fluxing away your furnace lining. You must always bear in mind your cost per pound of melted metal is the cost of the metal, plus the cost of the fuel to melt it, plus metal loss, plus refractory or crucible life.

For example, let's say we have a small rotary furnace and it is lined with x type of brick or x type of rammed monolithic lining. The lining cost us $500 in material (excluding labor). We must melt a total of 5000 pounds of brass with this lining and then it is shot, the flux we used has fluxed off the lining so bad that, with each heat, we were pouring a goodly portion of the lining off in the slag.

Disregarding everything else, our 5000 pounds of metal cost us the original cost plus $500. If we added the cost (which we must) to our metal cost, we are surprised at our cost. I have seen operations that were so bad that they were losing money on every casting.

What do you do? You can weigh the flux charge in and weigh the slag out, and the difference is lining and/or dirt. It doesn't take long to realize that if you are gaining great quantities of slag—and the inside diameter of your rotary is growing in leaps and bounds per heat—that your cover flux is doing something other than covering the melt.

Your several choices include adjusting your cover flux composition, using less flux, changing refractory composition, or all of the preceding. Correcting the problem by juggling these factors cannot only drive you nuts but be extremely expensive. So what's the answer? The best way to explain the correct way to go is for me to cite an actual experience I had.

Many years ago I took the job of superintendent of a large plumbing-ware foundry that was far in the red. The metal melted was a semired brass. The melting was done in three U.S. Rotary oil-fired furnaces, each with a capacity of 3000 pounds per heat. The linings were a cheap, stiff mud brick (a medium-duty fire brick). The slag was 50 percent Rasorite and 50 percent ground flint (window glass).

The cost of lining the rotaries was considerable, and a lining lasted only 25 heats or 75,000 pounds of metal. With the down time on a furnace—to re-line plus the labor and refractory cost—it ran into a considerable amount of money. The loss in production was another factor. We needed the output of all three furnaces (most of the time). The flux we were using was doing a very good job on the lining for sure. In place of analyzing the slag and making a change—which would have been a trial-and-failure procedure anyway—I decided to leave the flux as is and look at the lining material. I went to a well-known refractory company and, in place of asking their advice, purchased a selection of 20 different refractory bricks. The bricks ranged from a barbecue brick to super-duty, pure alumina brick. I had acid bricks, basic bricks—you name it. The price range ran from 15 cents per brick to in excess of $2 per brick. And as many heat ranges.

I placed the bricks in a kiln and put a 5-ounce pile of flux on each brick. The kiln was fired to 2800 °F, held there for 12 hours, and then allowed to cool to room temperature. The bricks were removed one at a time and sawed through at the point of flux contact to examine the amount of penetration (fluxing) done by the flux. See Fig. 3-10.

We found with this test that while some bricks were useless one had practically no penetration (chemical attack). We repeated the test, but this time we placed a pile of the flux mixed with 2 ounces of brass borings. The results were the same. The brick that showed zero penetration was a high-fired, super-duty fire brick in the $1 each range. I ordered enough of these to line one rotary.

At this point, the refractory company asked me about the intended use for this brick. I told them the brick was for use in a brass melting oil fired rotary with a Rasorite/glass flux cover. They advised me that the brick of my choice wouldn't last five heats. To make it short, 500 heats later we had only lost about half of refractory, and we re-lined at 2500 heats. I'm sure we could have squeezed in a few more.

This test is applicable to any refractory (such as plastic brick, etc.). Simply ram samples and fire, and then make your flux penetration test. While I do not suggest you discount advice from

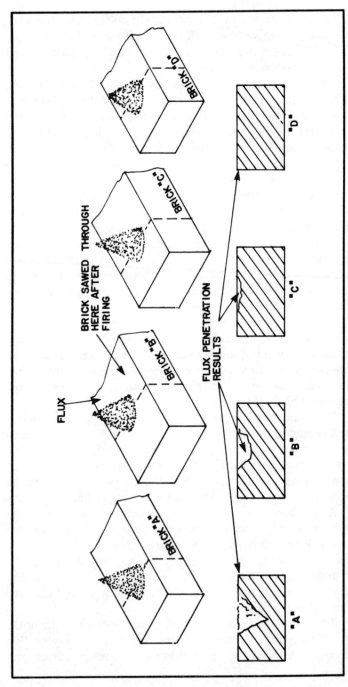

Fig. 3-10. The test method of selecting a refractory that is compatible with your flux and melting practice.

64

refractory companies, conducting your own testing is best.

DEOXIDIZING LEADED RED AND SEMI-RED BRASS

In the foundry we sometimes heard stories of mysterious substances added by the melter that would produce amazingly good results with a pot of red metal. These secret treatments were not magic; they were simply a good deoxidizer. Deoxidizers are often called scavengers or oxy-scavengers. The most used deoxidizers are phosphor-copper, phosphor-magnesium-copper and calcium boride. Of the three, 15 percent phosphor-copper (an alloy of 15 percent phosphor and 85 percent copper) is the most used. One ounce of 15 percent phosphor copper or the equivalent of 10 percent phos. copper is used to deoxidize 100 pounds of red metal.

This amount is equivalent to approximately 0.01 percent of phosphorus per 100 pounds of metal. The phos. copper can be added directly to the crucible. If the metal is to be received in a ladle, add it at this point. Skim the crucible or ladle full of metal and add the phos. copper by simply dropping it on the metal surface.

The reaction is very visible; the metal becomes fluid and bright. Allow a minute or two for the reaction, check the temperature, and pour. Weigh out your phos. copper each time; don't simply grab a hand full and toss it in. An excess of phos. copper will increase the fluidity of the metal to such an extent that the metal will actually penetrate the mold walls and cores, resulting in rough castings with sand inclusions. Excessive phosphorus will also result in dirty castings.

IMPURITIES

Free Iron. Free iron will cause random hard spots. If the iron is alloyed (in small amounts), it has little affect on the castings. For this reason it is permissable to use iron skimmers.

Sulfur. If kept to never more than 0.08 percent sulfur will cause no trouble.

Silicon. Silicon is real bad actor. As little as 0.01 percent in a leaded red brass will produce very unsound castings. It seems to go into solid solution and affects the crystallization. In addition, lead and silicon combine chemically to form glass. Silicon will cause the castings to be drossy, worm eaten, and have a whitewashed appearance.

Aluminum. Only a few thousandths of a percent of aluminum

65

in a leaded red brass or leaded semired brass can shoot the melt for you. Aluminum drastically alters the crystallization characteristics.

Magnesium. Having about the same effect on red brasses as aluminum and silicon, magnesium should be avoided.

By now you should be aware of the problems with melting scrap (especially mixed lots). You could have a nice pile of clean red brass valves, but what if only one of these valves has a silicon valve seat and you can't see it? You melt the batch and—bingo—you have one fine mess of unusable castings, plus a batch of metal you can't use. A silicon valve stem or nut, or a piece of aluminum scrap hidden in a red brass casting is all it takes.

If you are a secondary refiner or smelter, impurities are not a problem. You can take care of impurities by several methods; some are complex and some are simple, such as reducing the percent of the undesirable by dilution (thus reducing it to a level where it is no longer harmful). In a 100-pound pot of red brass, to have enough silicon to spoil the heat you are only looking at 0.11 ounce of silicon (eleven hundredths of an ounce or 0.01 percent). If you had 500 pounds, it would take .8 ounce of silicon to give you your 0.01 percent silicon. The smaller the heat the easier you can get into trouble.

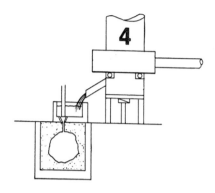

Leaded Yellow Brass

Leaded yellow brasses are a pleasing yellow color. They polish to a mirror finish, are easily machined, and are low in cost. Because they have only a moderate strength, they are not used for castings subject to severe engineering requirements.

When melting, gas absorption is usually no problem because the zinc vapor produced during melting continually purges, sweeping hydrogen away from the melt. It is used for ornaments, valves, lighting fixtures, light ferrules, bushings and plumbing ware, and builder's hardware.

Most foundries carry only one class of yellow brass (called No. 1 yellow or commercial No. 1 yellow). You have basically three yellow brasses: high-copper yellow brass, commercial no. 1 yellow, and yellow brass. See Table 4-1.

You will note that all three grades have only 1 percent tin. The reason for the tin is for strength. The lead is for machinability and ease of polishing.

FOUNDRY PRACTICE

The foundry practice for yellow brass (Table 4-2) is very much like leaded red brasses with some exceptions.

Scrap. Broken yellow brass scrap is easily identified by its color and fracture. Manganese bronze, which is actually a high strength yellow brass, is sometimes mistaken for leaded yellow. The fracture is different, however, and it is very much stronger

Table 4-1. Yellow Brass Properties.

Name & Composition	Properties	
High-Copper **Yellow Brass** 72% Cu 1% Sn 3% Pb 24% Sn	Weight lbs/in³ Patternmaker's shrinkage Solidification range Pouring temp. light work Pouring temp. heavy work Tensile strength Yield strength Elongation % in 2″ Brinell hardness (500 kg) Machinability Heat treatment	0.307 3/16% per ft. 1725-1700° F 1900-2100° F 1800-1900° F 38,000 PSI 13,000 PSI 35% 45 80 No response
No. 1 Yellow 67% Cu 1% Sn 3% Pb 29% Zn	Weight lbs/in³ Patternmaker's shrinkage Solidification range Pouring temp. light work Pouring temp. heavy work Tensile strength Yield Strength Elongation % in 2″ Brinell hardness (500 kg) Machinability Heat treatment	0.305 3/16″ per ft. 1725-1700° F 1900-2050° F 1750-1900° F 34,000 PSI 12,000 PSI 35% 50 80 No response
Yellow Brass 63% Cu 1% Sn 1% Pb 35% Zn	Weight lbs/in³ Patternmaker's shrinkage Solidification range Pouring temp. light work Pouring temp. heavy work Tensile strength Yield strength Elongation % in 2″ Brinell hardness (500 kg) Machinability Heat treatment	0.304 3/16″ per ft. 1725-1675° F 1900-2000° F 1750-1900° F 50,000 PSI 18,000 PSI 40% 75 80 No response

than leaded yellow. You soon become able to distinguish between the two. The only way to tell the difference between the three yellow brasses we have covered would be to run a simple copper-percent, quantititative-wet-chemical analysis. With mixed yellow

Ingot	#400	72-1-3-24
Ingot	#403	67-1-3-29
Ingot	#405.2	63-1-1-35

Table 4-2. Yellow Brass Alloy Forms.

scrap, it is not worth it to separate one from the other. Simply melt it as is.

In most cases the object itself is a give away. Small, cheap plumbing ware, "P" traps, bibs, etc., that are yellow in color surely would not be cast in manganese bronze, and nor would coat hooks, door knockers, or ornamental yellow castings. Yellow brass scrap (cast) should be purchased at a low price, and the scrap purchased should be as chunky as you can find.

Light, thin yellow brass scrap will produce a high melt loss due to zinc vaporization. This is a problem because most yellow brass castings are not heavy castings. Some red brass can be mixed in because the two are compatible. You can cut down your loss with light scrap by melting down a heat of heavy yellow and ducking the light material under to minimize its contact with the products of combustion. Of course you must watch out for random pieces of silicon bronze. Remember lead and silicon are not compatible.

MOLDING SAND

Most yellow brass is cast in natural bonded or synthetic green sand molds. The nature of the metal and the end use of the castings (cheap work) makes it too costly to cast in no-bake, dry-sand or core molds. You simply cannot get enough money for yellow brass castings to warrant a more expensive casting process. Green sand is the least expensive and most logical way to go. The pouring temperature is low enough that the refractiveness of the sand is no problem.

You need only a sand with a permeability of 20 and a green strength of 7 PSI with a maximum of 6 percent moisture. Actually, you can pour yellow brass castings up to 1/2 of an inch wall thickness and 50 pounds in weight before you need more permeability and green strength.

A typical natural-bonded sand suitable for most red and yellow brass castings would have about the following properties:

- ☐ Moisture: 6.7 percent
- ☐ Permeability: 12-20
- ☐ Green shear: 1.5 to 2.0
- ☐ Green compression: 6 to 8
- ☐ Clay substance: 10.2 to 14.5 percent

Yellow brasses have a tendency to dross so the molds must not be rammed too hard. If rammed too hard, the metal will boil against the hard spots, ruining the casting. Yellow brass is a much tougher metal to cast with good clean castings than red brass. It takes great skill in molding yellow brass.

FACING

It is not common to use any facing sands with yellow brass. A good fine grain molding sand will produce extremely smooth castings.

GATES AND RISERS

The gating and risering is very similar to red brass. An exception is that the gates and risers must be somewhat larger. Yellow brasses show fairly high shrinkage during solidification. Risers must be larger to provide for ample feeding.

Yellow brass must be gated in such a way that the mold must be filled as fast as possible. If slow pouring is practiced, zinc distillations will have time to form, and the casting will have a wormy appearance. These zinc tracks are caused by the zinc vapor rising from the metal as it enters the mold cavity and combining with the available oxygen to form zinc oxide.

One way that is practiced to prevent zinc distillation defects is to tip the molds up a few degrees to form a trap that prevents the incoming metal from sucking in air. See Fig. 4-1.

CORE PRACTICE

Cores for yellow brass should be free venting and at the same

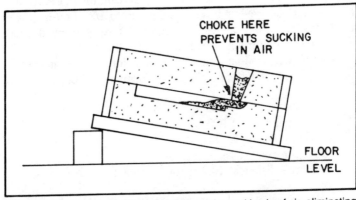

Fig. 4-1. Incline pouring of a mold to prevent the sucking in of air, eliminating oxidation problems and zinc distillation.

time not too open as to cause metal penetration that can be a problem with yellow brass due to its excessive fluidity. Straight oil sand cores with 1 to 2.5 percent of southern bentonite will prevent penetration. It is a general practice to coat the cores for yellow brass with a light core wash made with mica or graphite, or a commercial wash made of zirconium and graphite.

Cores must be hard and strong, but not too hard or strong or the result would be blows, hot tears (cracks), and problems in core knockout. In general, the core practice for red brass is applicable to yellow brass.

MELTING

It is best to melt yellow brass in crucible furnaces. With open-flame furnaces you get into very high zinc losses due to oxidation problems. With a rotary furnace, you need a foot of slag to prevent this.

In a reverberatory furnace the problem is accentuated due to the shallow bath with a large surface area. The general all-around pouring temperature is about 2050 °F. Anything above this results in heavy zinc losses and dirty castings resulting from the flaring zinc.

FLUXING AND DEOXIDIZING

Yellow brasses usually do not require a flux cover, nor any form of deoxidizing (in crucible melting). Gasses from the products of combustion are prevented by the surface vapor of zinc produced when melting sweeps the harmful gasses away (hydrogen, etc.). The yellow brass, when molten, is continuously purging itself.

With very thin castings to prevent misruns (cold shuts), a little phos. copper shot will increase the metal's fluidity. A small piece of pure aluminum will accomplish the same thing. The aluminum gives the castings a silvery surface appearance that some consider pleasing. Never use phos. copper and aluminum together. This will result in a very open and grainy structure and dirty castings.

I don't recommend either phos. copper or aluminum in yellow brass. If you use them, don't mix them; use phos. copper or aluminum.

The problem with the use of aluminum or phos. copper is that both are accumulative. The gates, risers, and scrap castings from a heat that was treated with aluminum or phos. copper, when melted again with new yellow, give you a heat that contains residual amounts of aluminum or phos. copper. (But how much is the question.) Then you add aluminum or phos. copper to this heat, and before long you are lost in the woods. In extreme cases, you would have to scrap the yellow brass because it would be so contaminated. You could only sell it as scrap to the junkyard. You don't want to borrow trouble. Melt in a crucible under a slightly oxidizing furnace atmosphere.

When you hit 2050 °F, skim and pour rapidly. Keep the ladle lip as close to the sprue as possible, and keep the sprue choked. Don't slack the stream, bobble, or dribble the metal in (pour hard).

If you like, you can use a thin cover flux of borax and flint glass while melting.

IMPURITIES

□ Iron causes grain growth, weak castings, and hard spots.

□ Phosphorus causes dirty castings if it is in excess of .015 percent.

□ Aluminum causes dirty castings and poor grain structure if it is in excess of .30 percent.

□ Silicon is a bad actor in leaded yellow brass (any leaded red metal).

□ Nickel is often included in amounts of up to .50 percent as a beneficial additive. It has a refining effect on the grain structure, producing a fine dense grain.

If you follow the rules of good foundry practice for yellow brass carefully without wide deviations, you can surely master it. When you get a process worked out, nail it to the wall.

Keeping records. Nail it to the wall is what we actually did years ago. With a gate of small castings we would be running again and again over the years, we would take a gate of these castings (like a bunch of grapes) of sprue, riser, gates and castings intact, and hang it on the wall with a nail. To this (sample) we would attach a record card with the metal composition, pouring weight, flask size, mold hardness, type of sand, core mix, yield weight (gates, sprues, runners minus finished casting weight) percent of scrap, defects, pouring temperature, etc. In a phrase, everything we needed to know about the job in order to repeat the job.

Before long you had a wall full of metal tied up. On large work or loose patterns that were hand gated, we would record all the information—gates, where located, size—and attach this record to the pattern. If the customer picked up his pattern, we would put the casting record in the files along with a description of the item, and customer's name etc. If a gated pattern or gated match plate came in, and the gating was wrong or would not produce the desired results for us, we would carefully remove the gates and regate the pattern. When the job was completed or pulled, we would remove our gating and replace the customer's gates as they were, saving the gating we found satisfactory or a precise record of the job which would allow us to repeat.

In some cases, a competitor's job would be pulled because of low price, only to return to us. A Polaroid camera is a must in the small jobbing foundry. Simply shoot several views and write the particulars, flask size, depth, etc., on the back of your picture records. If you made the original pattern equipment and core boxes, save the pattern layouts. If you are in the business to make money, keep good records of everything or you are not going to make it.

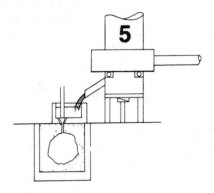

High-Strength Yellow Brass

High-strength yellow brass is called manganese bronze because of the small percent (3%) of manganese in the alloy. Even low-tensile yellow brass is called low-tensile manganese, and this alloy doesn't even contain manganese. It is a high-zinc alloy with 1 percent aluminum and 1 percent iron. You could lump manganese bronze and aluminum bronze together but I prefer to cover them separately.

High-strength yellow brass is simply an alloy of copper and zinc to which aluminum, manganese and iron is added (and in some cases nickel or tin). It is an extremely strong, tough yellow brass for use where high strength is required. In addition, there is excellent corrosion resistance, especially to saltwater. This accounts for its wide use in casting ship propellers. It is also widely used for nonsparking tools, gears, valve stems, cams, marine hardware, screw-down nuts, and machine parts.

Some sculptors call for their work to be cast in manganese bronze. The reason for this escapes me. The strength of manganese bronze comes primarily from the aluminum content that will run a range of from 0.75 percent to as high as 7.5 percent. The three most-used alloys run 1 percent Al, 4 percent Al and 6 percent Al, with a tensile strength range of 71,000 PSI, 95,000 PSI and 110,000 PSI. As the aluminum content increases the ductility decreases.

A great deal of emphasis has been laid on the difficulty of casting manganese bronze. Now it's not the easiest metal to cast, but a good foundryman who understands the metal can make excellent castings with low scrap loss. A great many founders—after just jumping in without understanding exactly the correct foundry practice necessary with manganese bronze—make a bust with it. They simply throw up their hands. You don't just jump in!

Let's talk about strength. A metal gets strength by consolidating its crystal structure into a very compact structure of fine crystals. A weak metal is one with coarse, loosely packed crystals. As with most things, a loosely woven material made of fibers with a loose cell structure is by far weaker than one of a tight weave made with a fiber that has a tight cellular structure.

In numbers there is strength. The metal with the most crystals per cubic inch is stronger than 1 cubic inch of a metal with much less crystal population (larger crystals). Disregarding the individual shape of the grains of sand, a fine sand will pack harder than a coarse sand.

Steel is stronger than iron, and manganese bronze is stronger than red brass. The faster a metal goes through its solidification range the smaller the crystal structure will be (thus more compact and stronger). In a metal that has a wide solidification range, the crystals form over a longer period of time and this results in the formation of larger crystals. We chill cast iron to form a compact, extremely hard iron of a close grain with a minimum of free graphite.

If you hammer on a piece of soft copper, you can increase its strength and hardness considerably; what you are doing here is work hardening. Driving the individual crystals of copper closer together makes the metal denser. If you anneal, soften the copper, you are loosening up the crystals.

Let's get back to the manganese bronze. With manganese bronze, you have a very short solidification range and thus considerable shrinkage.

Because you are dealing with a high-shrinkage metal, let us talk about shrinkage. Manganese bronze has a patternmaker's shrinkage of 1/4 of an inch per foot. This is the allowance you must add to the pattern to compensate from pattern actual size to casting size at room temperature. Now don't confuse this shrinkage with volumetric and solidification shrinkage. These two shrinkages are the ones we are primarily interested in when we consider the feed metal (risers). We have to supply liquid metal to the casting as it solidifies to take the place of the voids as liquid metal solidifies and becomes solid.

All metal when cast goes through three distinctive separate phases of shrinkage.

☐ The first phase is a volumetric shrinkage. This is a reduction of the metal as it cools from the pouring temperature to the point where it starts to solidify. If we pour a low-tensile strength manganese bronze at 2000 °F, it will remain liquid from our 2000 °F temperature to 1616 °F—a range of 384 °F. When you reach 1616 °F, this is where the solidification starts. During this range, you have no real physical change going on other than a reduction in volume.

☐ Phase two is solidification shrinkage. This is a reduction in volume as the metal passes from a liquid to a solid. This is also when the liquid metal crystallizes. With our low-tensile strength manganese bronze, the shrinkage and physical change is called the solidification range. It starts at 1616 °F and stops at 1563 °F for a temperature (span) or range of only 53 °. This is not only a short range, but this 53 ° is lost very fast due to the mold walls rapidly absorbing heat from the metal. This accounts for the rapid formation of close-knit fine crystals that makes for a dense metal structure of high strength.

Now let's compare this with 85,5,5,5 red brass. The solidification range of red brass is from 1850 °F to 1570 °F (or a 280 ° spread). Under the same cooling conditions (rate of heat absorption from the metal), red brass will take considerably longer in time units to completely solidify than manganese bronze, producing a coarser less compact weaker metal structure.

☐ With phase three, you have contraction shrinkage. This is the reduction in size of a casting as it cools from the solidification temperature to room temperature.

The rate of solidification (crystallization) is directly proportional to the ability of the mold material to carry away the heat. It is also directly proportional to the area of metal to sand (mold material), contact, and it is inversely proportional to the volume of the casting. Naturally, the bigger the casting the more heat reduction.

Because manganese bronze has a high shrinkage rate, large risers must be used so that all parts of the casting are adequately fed during the #2 solidification range. Chills can be used to an advantage to reduce the size and number of risers.

Insulating sleeves and hot-topping compounds will help in making the risers feed more efficiently. Liquid feed metal must be supplied to the casting until solidification is complete (crystallization).

Note: All the compositions for various alloys in this book are typical. They all vary somewhat. For example I give a typical composition of a low tensile manganese bronze as copper 58 percent, tin .5 percent, zinc 39.5 percent, iron 1 percent, aluminum 1 percent with 0 percent manganese. You will run into low-tensile strength ingot or scrap that might analyze as copper 58 percent, tin 0 percent, zinc 39 percent, iron 1 percent, and manganese 1 per-

Table 5-1. Manganese Bronzes.

High Strength Yellow Brass Low, Medium, and High-Tensile Strength	
Name & Composition	**Properties**
Low-Tensile Manganese Bronze 58% Cu .5% Sn 39.5% Zn 1.0% Fe 1.0% Al	Weight, lbs/in³ 0.301 Patternmaker's shrinkage 1/4″ per ft. Solidification range 1616-1583° F Pouring temp. light work 1900-2000° F Pouring temp. heavy work 1750-1900° F Tensile strength 71,000 PSI Yield strength 28,000 PSI Elongation % in 2″ 30% Brinell hardness (500 kg) 100 Machinability 26 Heat treatment No response
Medium-Tensile Manganese Bronze 64% Cu 26% Sn 3% Fe 4% Al 3% Mn	Weight lbs/in³ 0.288 Patternmaker's shrinkage 1/4″ per ft. Solidification range 1725-1650° F Pouring temp. light work 1900-2000° F Pouring temp. heavy work 1800-1900° F Tensile strength 95,000 PSI Yield strength 48,000 PSI Elongation % in 2″ 20% Brinell hardness (3000 kg) 180 Machinability 30 Heat treatment No Response
High-Tensile Manganese Bronze 63% Cu 25% Zn 3% Fe 6% Al 3% Mn	Weight lbs/in³ 0.283 Patternmaker's shrinkage 1/4″ per ft. Solidification range 1693-1625° F Pouring temp. light work 1950-2150° F Pouring temp. heavy work 1800-1950° F Tensile strength 119,000 PSI Yield strength 83,000 PSI Elongation % in 2″ 18% Brinell hardness (3000 kg) 225 Machinability 8 Heat treatment No response

cent. All ingot and metal specifications have a leeway plus and minus or a minimum and maximum as to the various components and still fall into the ballpark. See Table 5-1.

Let's look at two commercial manganese bronze ingots available and the leeway allowed by the ingot makers (smelters) as to chemical composition. See Table 5-2.

Physical properties are usually presented as minimum properties (such as tensile strength, yield strength, elongation, Brinell hardness, reduction of elasticity, compression strength). If the manufacturer of the ingot says that his ingot #421 (or what have you) has a tensile strength of 71,000 pounds per square inch, he is giving the minimum you should expect. It could be stronger but should not fall below 71,000 PSI. He might show on the chemical specs for a given heat of ingots that 2 percent lead is the maximum lead allowed for this composition. This does not mean that the lead was intentionally added up to 2 percent. What he means is that the alloy will tolerate up to 2 percent lead without materially affecting the physical limits for the alloy. You can pick up that much lead in the ordinary course of foundry practice.

Because all the physical properties can vary due to foundry practice (melting, gating, etc.), when you melt and cast from an ingot or ingots that the smelter has designated as having a minimum strength, elongation etc., this is no guarantee that the castings made from this metal will have the physical properties. This lets him off

Table 5-2. Alloys.

Alloy #1 (low tensile)	Copper	Min.	56%	Iron	Min.	0%
		Max.	62%		Max.	2%
	Manganese	Min.	0%	Tin	Max.	1.50%
		Max.	1.5%	Lead	Min.	0.50%
	Aluminum	Min.	0%		Max.	1.50%
		Max.	1.5%	Zinc	Remainder	
Total of other constituents Max. 0.25%						
Alloy #2 (high tensile)	Copper	Min.	60%	Iron	Min.	2.0%
		Max.	68%		Max.	4.0%
	Manganese	Min.	2.5%	Tin	Max.	0.5%
		Max.	5.0%	Lead	Max.	0.20%
	Aluminum	Min.	3.0%	Zinc	Remainder	
		Max.	7.5%			

of the hook (and rightly so) because the ingot maker is only giving you the properties of the ingot he sold you based on his testing. He has no control over how you use his product (or misuse it).

Should you, when melting, overheat and thereby flare off the zinc in considerable amounts, this zinc loss changes the alloy and will materially alter its physical properties. If you fail to supply sufficient feed metal (risering, etc.) and wind up with a casting of coarse structure or shrinkage porosity, you can't blame this on the ingot supplier. The only way you can nail him is if you run chemical and physical tests on his ingot and find that he has misrepresented the material.

ALLOY FORM

Use ingots or scrap to make your own manganese bronze. You can purchase ingots from a reliable supplier. Three typical ingots are Ingot #421, low-tensile strength, minimum 71,000 PSI; ingot #423, medium-tensile strength, minimum 95,000 PSI; ingot #424, high-tensile strength, minimum 119,000 PSI. Consult your metal supplier. With scrap, you are never sure of what you are getting unless you have a lab to check the composition.

Making your own? I made lots of manganese bronze from virgin metals when it was not as available as it is today, and when I could not wait for the ingots. The procedure is quite simple. What we did was melt up what is known as a hardener that we cast into very thin ingots. The hardener was an alloy made by melting the manganese, aluminum, and iron in just enough copper to give us a castable hardener.

When we wanted a batch of manganese bronze we would compute, by percentages, exactly how much hardener, zinc, and copper we needed for the class of manganese bronze we wanted. We melted down #1 copper scrap or ingot copper, and added the percent of hardener based on our calculations. When the hardener was completely in solution with the copper, we then added preheated zinc in small pieces to bring the zinc percent up to our desired alloy composition. We then stirred the metal to make sure of a good physical and chemical mix and cast the melt into ingots (which we remelted for casting).

Hardeners can be purchased from most secondary refiners; this eliminates computing and making them. They are fairly foolproof—with directions of exactly how much hardener to use per 100 pounds of copper—to produce any range of tensile strength. I don't think they are referred to anymore as hardeners; they are now called master alloys. Master alloys are available in hundreds of combinations and specifications for just about any use you could think of for ferrous and nonferrous work. They are used to make up alloys and adjust alloys.

The old saying was the business of the foundry was making castings not metal. That was good advice when you could deal with the ingot makers directly at a reasonable price. This is now getting difficult and extremely costly, and you have several middlemen between you and the smelter, each one with his hand out.

In order to make ingots to specification, whether you use primary metals or secondary metals, you must have some chemical control. It need not be complicated or extensive, but you cannot it without the minimum lab equipment and know how to run both quantitative and qualitative chemical tests.

MOLDING SAND

Most molding sands used to cast red or yellow brass castings are usually suitable for manganese bronzes. The aluminum in the alloy forms a thin, protective film on the metal during casting. This film is the aluminum oxide Al_2O_3 called alumina, and it gives the casting a nice smooth finish even when casting in a fairly open coarse grained sand. See Fig. 5-1.

Because manganese bronze is cast over a wide weight range—from as light as a few ounces to in excess of 5 tons—such as a large ship propeller, you have to increase the green compression strength and permeability as you move up to heavier work. The general recommendations for green sand specifications are as follows:

☐ Light work weighing under 100 pounds with a section thickness of 3/4 of an inch maximum—permeability 20, and green compression strength 8 PSI.

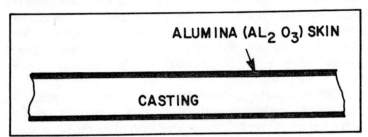

Fig. 5-1. Cross section of a casting showing a skin coating of alumina Al_2O_3.

☐ Light-to-medium work weighing from 100 pounds to 250 pounds, with a section thickness of 1 1/4 inch maximum permeability 40, and green compression strength 8 PSI.

☐ Above 250 pounds, I would move to a no-bake system such as CO_2 or furan binders for safety sake. Before no-bake binders it was common to pour extremely large manganese bronze castings in green sand.

☐ For 500 to 1000 pounds, with a section thickness from 1 3/4 to 2 1/2 inches, the permeability should be 100 and the green strength should be 12 PSI. From here to 5 tons or more with a section thickness running from 2 1/2 inches and upward, you are looking at a permeability of 150 plus and a green strength of 15 PSI plus.

Refractiveness of the sand is not important due to the low pouring temperature. A good, general synthetic sand for manganese bronze is a simple mixture of washed and dried silica sand, 100 mesh, bonded with 4 to 4.5 percent southern bentonite, and tempered with 4 to 4.5 percent moisture. Green strength is 6 to 7 PSI and permeability is 30 to 40 (up to 150 pounds in weight).

For heavier work, simply increase the permeability by moving to a coarser sand and add bentonite to increase the green strength. I usually add 1/2 percent to 3/4 percent fine wood flour. It is not recommended but up to 3.5 percent of sea coal works wonders. Albany natural bonded sands (grade 0 to 1 1/2) are excellent for light to medium work.

FACING

Due to the alumina skin along with the low pouring temperature, facing is not required.

GATES AND RISERS

Pouring manganese bronze is like pouring soap and water. Any agitation, swirling, nozzling, or squirting will result in the formation of dross, resulting in a scrapped casting. Under no conditions can you pour manganese bronze directly into a mold through a simple down gate or through a riser. The metal must fill the mold by simple nonagitated displacement from the bottom of the mold cavity.

Let's nail this point down. "Agitation" results in dross, bubbles, froth—which equals big trouble. See Fig. 5-2.

As shown in Fig. 5-2, you have the simplest arrangement of

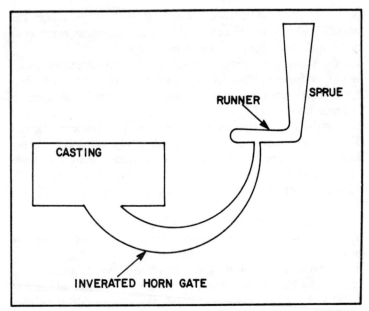

Fig. 5-2. Cross section through a casting gated by a simple inverted-horn gating system.

displacement gating using an inverted horn gate. The horn gate is invaluable in gating manganese bronze. This gate is a horn-shaped set gate and is rammed into place when ramming the drag flask. See Fig. 5-3.

For the steps in using a horn gate, see Fig. 5-4. Figure 5-4 shows

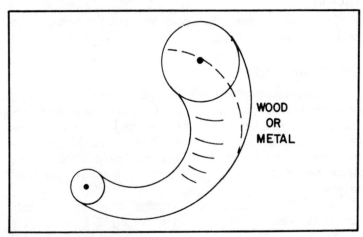

Fig. 5-3. A horn gate pattern.

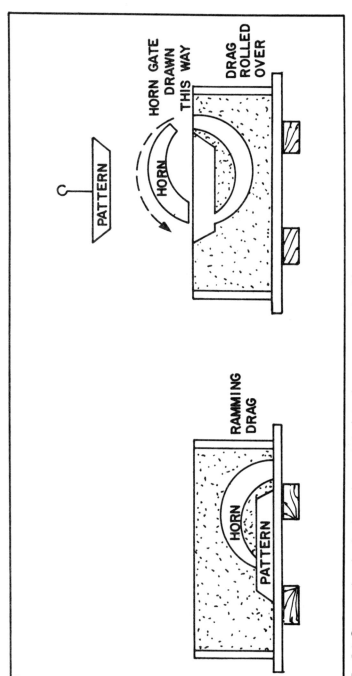

Fig. 5-4. Cross section showing how a horn gate is drawn from a green sand mold.

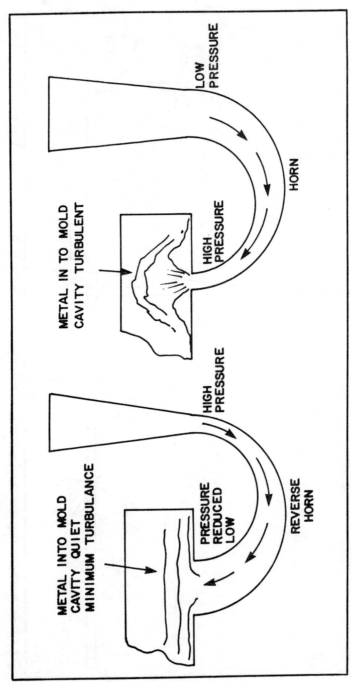

LOW PRESSURE

HORN

HIGH PRESSURE

METAL IN TO MOLD CAVITY TURBULENT

HIGH PRESSURE

REVERSE HORN

PRESSURE REDUCED LOW

METAL INTO MOLD CAVITY QUIET MINIMUM TURBULANCE

Fig. 5-5. Pressure relationship with the large end of horn against pattern Vs, the small end.

the normal method of using a horn gate (with the small end against the pattern).

Using the horn this way increases the chance of nozzling the metal in the mold. That is fine for most metals but a definite no for manganese bronze. We reverse the horn when gating manganese bronze—placing the large end against the pattern that has the effect of reducing the metal velocity into the mold cavity, thus minimizing the chance of causing turbulence, churning or nozzling—which results in drossing. The method of gating is referred to as reverse horn gating. See Fig. 5-5.

Often a loose pattern for manganese bronze is marked with a circle in which the letters RHG are painted. See Fig. 5-6. This simply means reverse horn gate for this pattern.

Horn gates can be made in the form of a dry sand core (hollow horn) and rammed up in place and left there. This method is used where a loose horn pattern would be difficult to draw or impossible to draw. It is made in two halves from a left- and right-hand core box and then glued together. See Fig. 5-7.

This arrangement makes it possible to get in close to the job or increase the reach. That could be a problem with a horn that

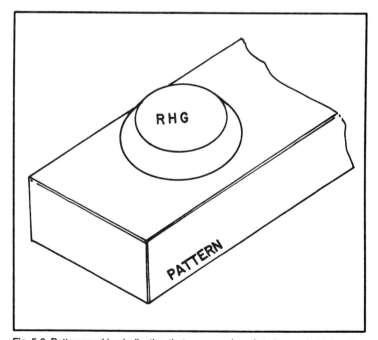

Fig. 5-6. Pattern marking indicating that a reverse horn is to be used at this point.

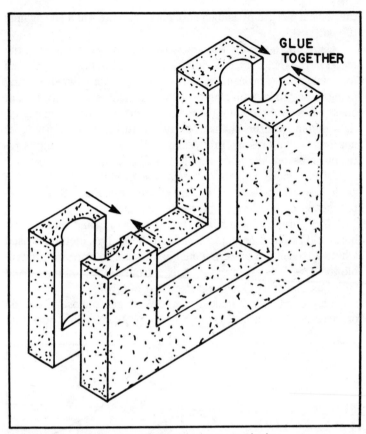

Fig. 5-7. A horn gate made from a baked core sand mix.

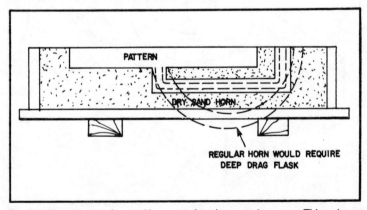

Fig. 5-8. How to use a dry-sand horn gate for a long reach or span. This reduces the depth of cope required when using a conventional horn pattern.

must be drawn. Therefore it must be a part of a circle so that it can be drawn. See Fig. 5-8.

Another method of horn gating is by putting the horn in the drag with the pattern in a cheek and cope. With this method, the gate can be cut by hand or done with set gates. See Fig. 5-9. This method is widely used when molding medium or large work. You really don't have a choice; you must get the metal into the mold with no agitation. Now let's look at the principle design.

Figure 5-10 shows the proper method for gating a medium or large propeller, a small propeller, a large or medium gear blank and a small gear blank.

In many cases you can use a physical skimmer to hold back pools of floating dross (should they form), and to prevent them from becoming washed into the mold cavity proper. I have poured a considerable number of large manganese bronze castings, and I have used this method with great success. A sprue large enough to introduce a steel-bladed paddle welded to a thin rod is cut about midway between the pouring sprue and the ingate.

When the casting is being poured, a molder or helper catches any floating dross with the paddle, and holds it back until the pouring is complete. It's tricky to do and takes practice. See Fig. 5-11.

Another method of trapping any dross floaters is the use of nails to catch the dross (like the grate bars of a catch basin.) See Fig.

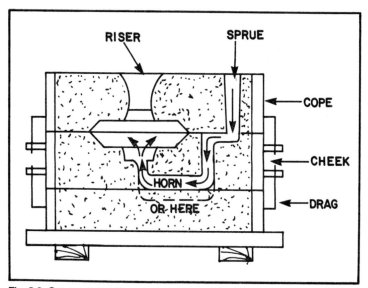

Fig. 5-9. Cut green sand horn gating accomplished by using a three-part flask.

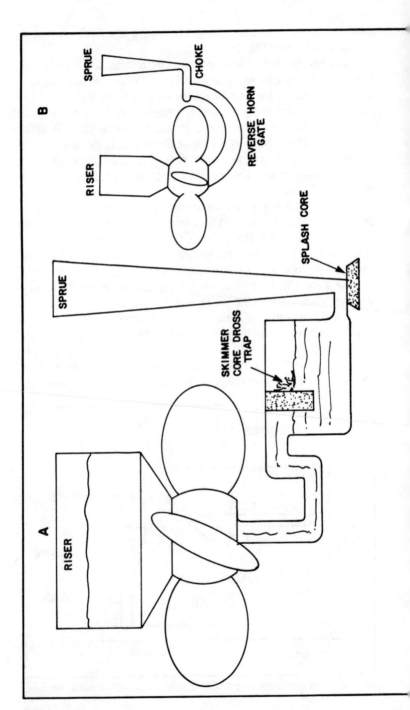

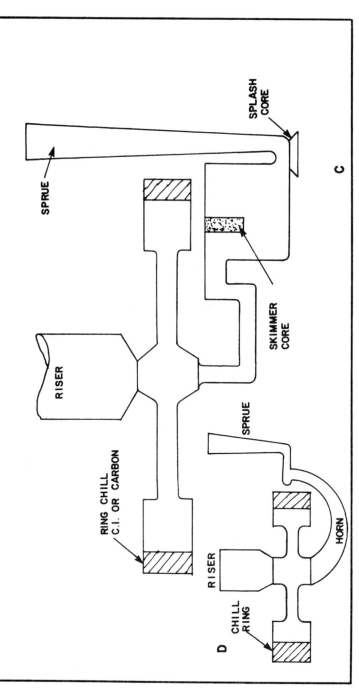

Fig. 5-10. Typical gating arrangement used to cast a large propeller (A); gating a small propeller (B); gating a large gear blank (C); gating a small gear blank (D).

SPRUE

SPLASH CORE

C

SKIMMER CORE

RISER

RING CHILL C.I. OR CARBON

SPRUE

HORN

RISER

CHILL RING

D

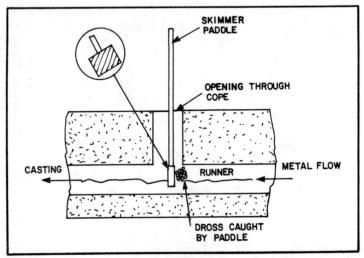

Fig. 5-11. Catching dross manually through a cope opening with a steel paddle.

5-12. Even with an elaborate gate (as shown in the large propeller, Fig. 5-10), you could, for safety sake, put a nail skimmer across.

Strainer cores are never used in the gating system for manganese bronze because they only aggravate the condition and

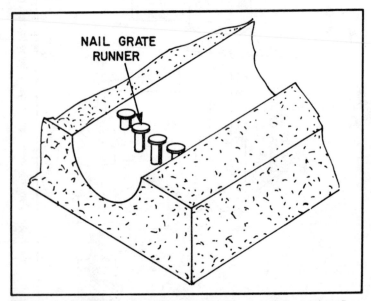

Fig. 5-12. Nails across a runner or gate used to trap dross to keep it from flowing into casting cavity.

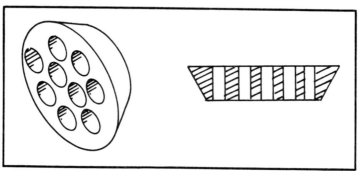

Fig. 5-13. A view and cross section of a skimmer or strainer core.

produce drossing. See Fig. 5-13.

With small multiple patterns on match plate, you are much better off to come into the runner with a little reverse horn gate, have a runner bar in the drag as well as the cope, and slant the gates back so that the system fills up (the runners). Then the metal backs up into the mold cavities quietly with a minimum of turbulence. Make sure that the cope runner bar is large enough so that it is the last thing to solidify and close enough to the gates to keep them open so the bar can supply liquid feed metal as the casting shrinks. See Fig. 5-14.

If the gate between the runner and the casting is too small or too long, it will freeze before the casting and cancel out the runner and its purpose. You must always have progressive solidification. See Fig. 5-15.

The use of a runner bar in the drag is to provide more bulk (heat) to the feed system and to help minimize turbulence. It need not be very large because it plays no part in supplying feed metal. It will help keep the runner liquid longer so that progressive solidification can take place. See Fig. 5-16.

I have said a lot about dross because its inclusion in a casting usually results in a scrap casting or at best one very dirty, scabby casting. The generation of dross in manganese bronze is caused by only one thing, and that's the agitation of the liquid metal. If you were to take a crucible full of molten manganese bronze, you could stir or agitate it enough with a skimmer that the entire contents is reduced to a big blob of nothing but dross. The dross produced by improper gating or pouring usually floats on the top of the stream and resembles beer foam.

Dross defects in castings are usually found on the cope surface of the casting as an ugly, spongy multicolored area of loose-knit

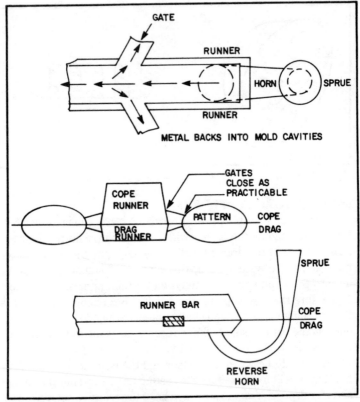

Fig. 5-14. Runner bar and gate construction for plated patterns used to decrease turbulence by arresting the flow of incoming metal then backing the metal into the mold cavities.

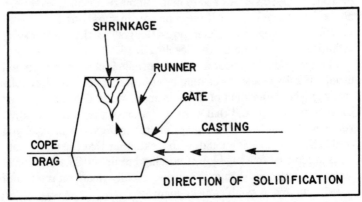

Fig. 5-15. On a plated pattern, the runner must be large enough to confine the shrinkage to the runner.

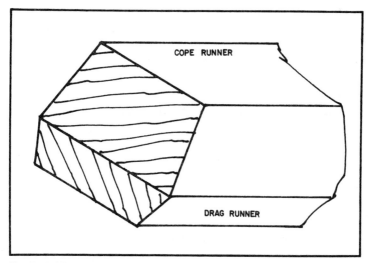

Fig. 5-16. Visual ratio of cope runner to drag runner in size.

stringy oxides of copper, aluminum, and manganese oxide. When you pick at it with a knife point, it will flake away easily. The more you dig the deeper it can get. In some examples, I have seen the dross go completely through the casting.

Dross does not always make its presence known by showing on the cope surface. It can be an internal defect that makes itself known only by x-ray, during machining, or when the casting fails in service. This last one can lead to a bad accident and/or a huge product liability suit. See Fig. 5-17.

Now let's look at risers. Manganese bronze flows into the mold with great ease. This is a plus. If agitated it will produce dross.

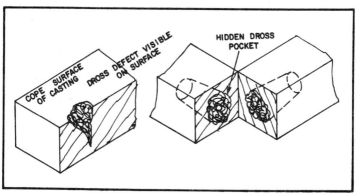

Fig. 5-17. Illustration of dross inclusions in and on a casting.

Now let's say you have been successful in getting the mold filled with a minimum of agitation. Using good gating practice, what do you do about its high shrinkage characteristic. When the mold is full and the metal starts to cool and shrink, certain portions will cool and shrink faster than other portions. This is due to design (varying metal section). When this happens, the section that is cooling ahead of others will draw liquid feed metal from the nearest available source of liquid metal. This causes a shrink.

I seem to harp on directional solidification but it is a must. Let us look at the problem from a different angle. Let's consider the gates, runners, risers, and the casting as a single unit. You know that you are going to have shrinkage so all you have to do is, by directional solidification, restrict the solidification problems to the risers, runners, and sprues. In doing this, you have confined the problem to a part or parts of the whole that you are going to remove anyway.

Figure 5-18 shows a simple tapered wedge-shaped casting. The heat will be lost fastest in the thinner section. The solidification is directional from thin to heavy. Therefore, to feed this solidifying section the liquid phase moves from heavy to thin. As shown in Fig. 5-18, the gate is used only to fill the cavity. Because it is too thin and freezes first, the mold is filled with liquid metal having no means of supplying feed metal other than from itself. The resulting casting will have a shrink cavity in the heavy section

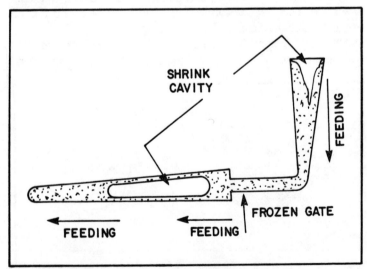

Fig. 5-18. The effect of a long, thin gate results in shrinkage.

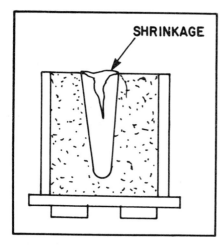

Fig. 5-19. A gateless casting slows shrinkage.

because, when it finally solidified, there was no supply of feed metal available to hold it up. See Fig. 5-19.

By supplying a reservoir of liquid metal with sufficient volume, it will remain liquid and feed the casting. Because you are never going to get rid of the shrink, what you do is simply move it away from the casting.

Figure 5-20 shows the wedge with a riser attached to feed the casting as it solidifies. Have you lost the shrinkage itself? No, you simply moved it back into the riser; it is still with you but now it's where you want it. When we degate the casting, the shrink makes the next heat of metal.

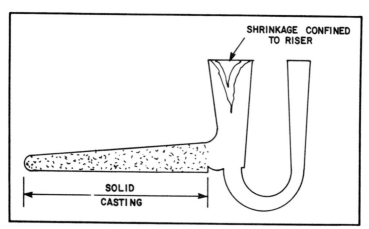

Fig. 5-20. The purpose of the riser is to supply liquid feed metal promoting directional solidification.

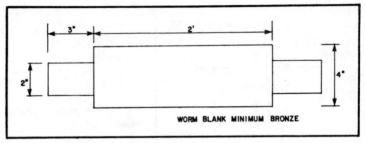

Fig. 5-21. A blueprint for a worm gear blank.

Let's look at a really tough job to feed in manganese bronze. You have to cast a large blank for a worm gear for a heavyduty speed reducer, hoist, or perhaps draw works of some sort. The casting desired is as shown in Fig. 5-21.

This doesn't look tough to cast in manganese bronze. Let's say you make a split pattern and cast it horizontally. Because the casting is rather heavy, approximately 101 pounds, you will have to apply a riser of considerable size to take care of shrinkage. You could put it in the center. See Fig. 5-22.

You might be able to hold it up with this arrangement and may-

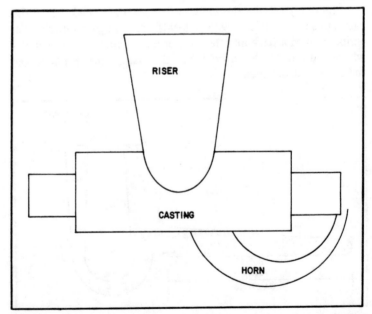

Fig. 5-22. Proposed gating and risering for the worm gear blank shown in Fig. 5-21.

96

be not. The point of riser attachment is going to be as large as the maximum diameter of the casting (plus with very large fillets). This represents a considerable riser removal problem, and it is extremely costly.

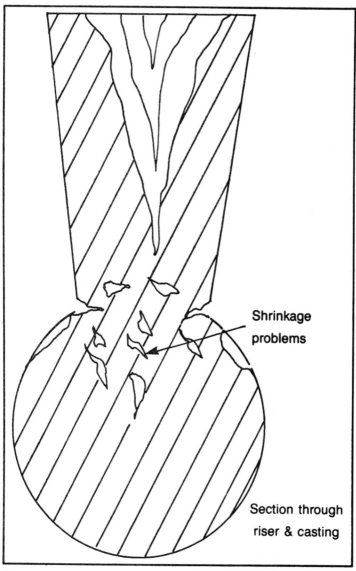

Fig. 5-23. Cross section of probable defects you would have by gating worm gear as shown in Fig. 5-22.

Should you have an internal shrinkage under the riser after working like a beaver to remove the riser—only to find you have a scrapper—you could have in the area of the riser some shrinkage porosity that would make the worm (after machining) unsound. The result would be a mechanical failure along the line. See Fig. 5-23.

You could elect to riser on one end. See Fig. 5-24. Here you have the same problem as with central risering. So where do you go?

Because you are going to have to have a large riser for feed that must be removed anyway, the most logical way to cast your worm blank is to make the riser part of the pattern. Then you can saw the riser off and turn the worm blank on the lathe. You have to turn the blank anyway. See Fig. 5-25.

You can cast bushings or bar stock horizontal with good results by pouring with the riser 1 inch to the foot lower than the sprue. When metal shows in the riser above the casting, the sprue is frozen off with wet molding sand and an ingot. The mold is then tipped the other way, and the riser is filled with hot metal to feed the casting. See Fig. 5-26. The same gating and pouring arrangement is used for pieces of flat and square stock. See Fig. 5-27.

Let's take a case where you have a disc shaped casting with a hub. See Fig. 5-28. Say you simply gated it with a horn on the hub with no riser. You will get results as shown in Fig. 5-29.

Now you have two ways to correct this shrinkage problem. One way would be to apply a chill on each side of the hub. See Fig. 5-30. The purpose of the chill, in this case, is to try to suck enough heat

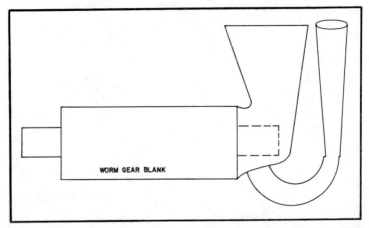

Fig. 5-24. An alternate gating system for our worm gear (Fig. 5-21) that would probably produce a defective casting no better than the gating shown in Fig. 5-22.

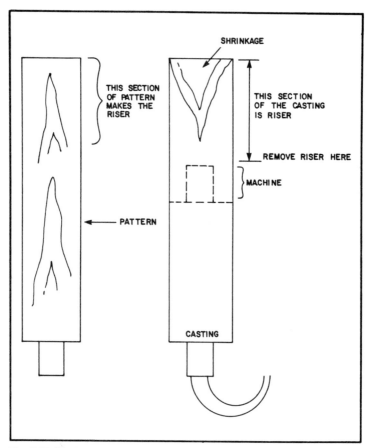

SHRINKAGE

THIS SECTION
OF PATTERN
MAKES THE
RISER

THIS SECTION
OF THE CASTING
IS RISER

REMOVE RISER HERE

MACHINE

PATTERN

CASTING

Fig. 5-25. The best solution to casting the worm gear shown in Fig. 5-21 and proposed pattern change.

out of the hub so that it solidifies at the same rate as the web for one-directional solidification toward the riser that will supply the feed metal to the casting and move the hub shrink cavity into the riser out of the way. See Fig. 5-31.

It would be nice if you could do that but it is impossible, you would have to precisely control the heat absorption of the chills so that it all comes together at the precise time.

If you didn't chill it enough, you would still wind up with a shrinkage defect in the hub. If you chilled it too fast, you would have shrinkage cavities on each side of the hub. See Fig. 5-32.

The solidification progress is in one direction and can be fed with liquid metal that is to come into the hub with a horn gate and

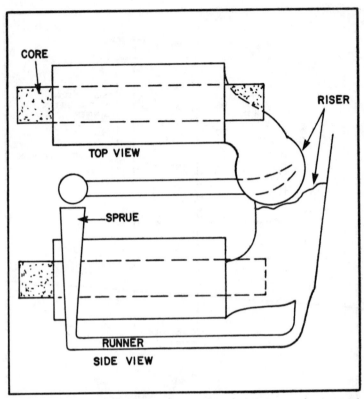

Fig. 5-26. A good gating system for bushings in manganese bronze cast horizontally.

have a large enough riser that it will solidify last and with sufficient volume to feed the casting. See Fig. 5-33.

Remember that the feed riser must be large enough with sufficient volume to solidify last, and restrict the shrinkage to the riser. Too small of a riser increases the problem and the defect. It is usually much worse than if it had been left off all together. Also a riser that is not fed hot metal from the gating system is useless. See Fig. 5-34.

I don't know how many times I have seen a casting spoiled due to a riser being set where none was needed, the riser was too small to be effective,or the gating system was not designed to supply the riser with hot metal. With cast iron where you have a very broad solidification range and low shrinkage, it is very common to pour the casting through the riser (therefore assuring hot feed metal in the riser). See Fig. 5-35.

100

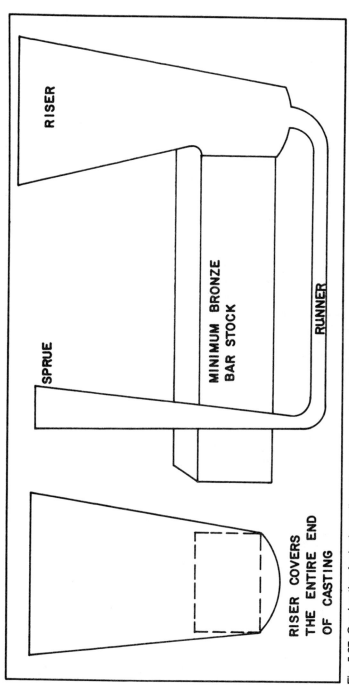

RISER

SPRUE

MINIMUM BRONZE
BAR STOCK

RUNNER

RISER COVERS
THE ENTIRE END
OF CASTING

Fig. 5-27. Good gating for horizontally cast manganese bronze bar stock.

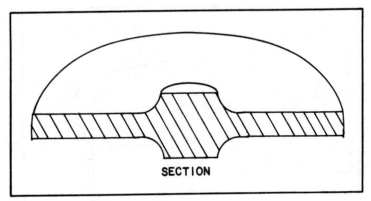

Fig. 5-28. Blueprint of hubbed disc.

Cast iron doesn't foam. Try this with manganese bronze and you will only do this once.

The yield in casting weight against pouring weight (risers, gates, etc.) for manganese bronze is very poor. It is common to have a 20-pound valve casting in manganese bronze have a 40-pound-plus riser (giving you a negative yield).

Let's look at a casting for a bell crank where you have a pad or heavy bearing section on the ends and a pivot point hub (Fig.

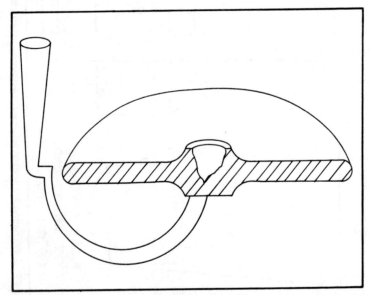

Fig. 5-29. Casting shown in Fig. 5-28 cast without a riser shrinks badly at the hub.

102

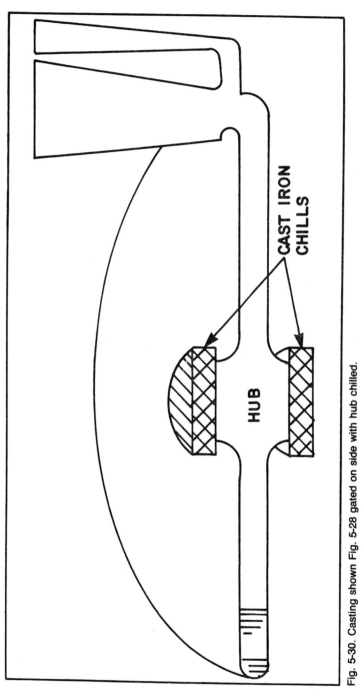

CAST IRON CHILLS

HUB

Fig. 5-30. Casting shown Fig. 5-28 gated on side with hub chilled.

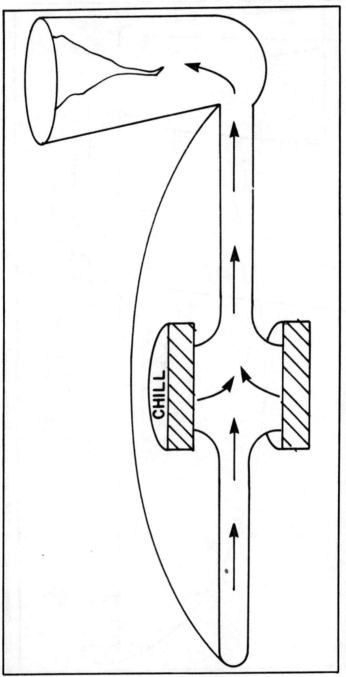

Fig. 5-31. The side gate and precise chilling as shown probably will not work.

104

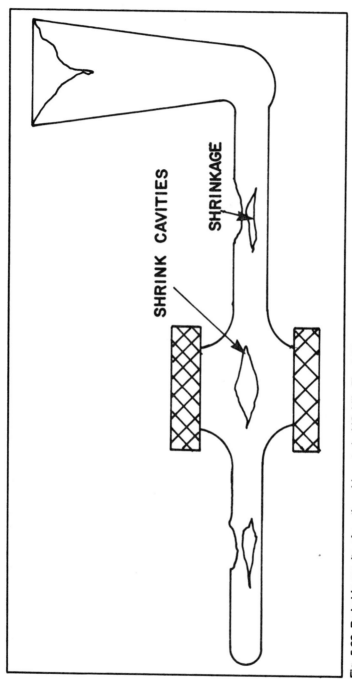

SHRINK CAVITIES

SHRINKAGE

Fig. 5-32. Probable results of casting side gated with chills Fig. 5-31.

105

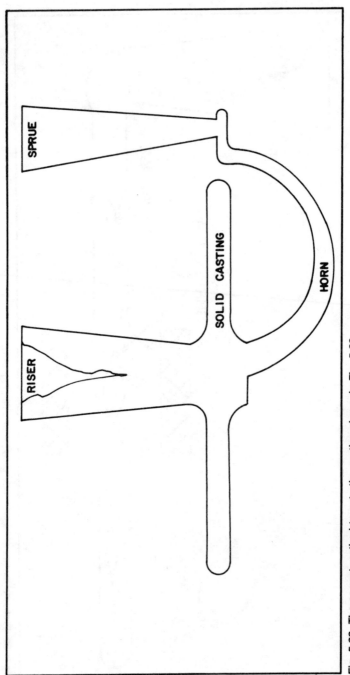

Fig. 5-33. The correct method to gate the casting shown in Fig. 5-28.

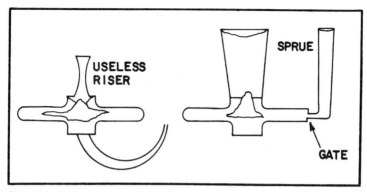

Fig. 5-34. Two common errors in gating.

5-36). Here is a classic case where chills would do the trick. See Fig. 5-37.

If you elected to not chill the two bearing sections on the ends, then you would have to riser them and gate into all three risers. See Fig. 5-38.

Chills are to be avoided rather than sought. When there is no other way of feeding an isolated section, you have little choice. If

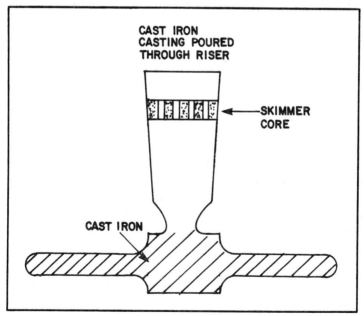

Fig. 5-35. It is a lot simpler if casting shown in Fig. 5-28 is cast in gray iron instead of red metal.

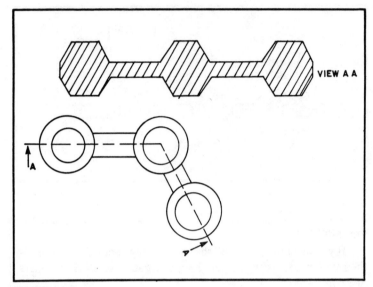

Fig. 5-36. Blueprint for a bell crank.

improperly placed, they will do more harm than good. You must locate them in such a manner that the inaccessible area chilled will solidify well before the neighboring portions of the casting (which can later draw upon the risers).

Chills are usually cast iron and must be clean, dry and free from rust. A thin coating of lube oil and graphite will keep them from blowing where they contact the casting. You can also coat them with linseed oil, roll them in fine, sharp sand, and oven dry them to set the linseed oil. See Fig. 5-39.

With all gating—regardless of the metal or alloy being cast—the same rules apply. The mold must be filled as fast as possible with a minimum of turbulence and gated in such a manner as to affect progressive solidification.

Gating is a common-sense proposition if you know your alloys, characteristics, solidification range, total shrinkage, fluidity, problems with drossing, and zinc distillation. By studying the design of the casting, you simply have to put it all together. There are various formulas for gating that give various ratios of gate, runner, riser, and size. That is all fine and good, but as every casting is designed differently you really cannot boil it down to a fixed mathematical formula.

Riser can be insulated with sleeves or covered over with charcoal, vermiculite or other insulators or exothermic compounds. Ex-

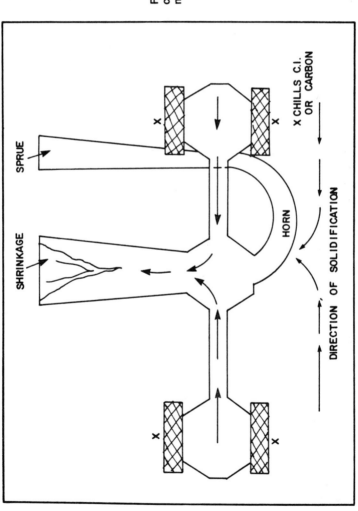

Fig. 5-37. Correct gating for bell crank shown in Fig. 5-36, if cast in manganese bronze.

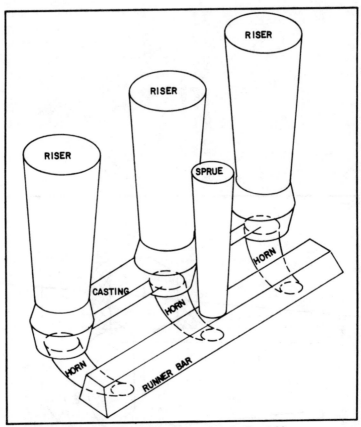

Fig. 5-38. Alternate gating for bell crank shown in Fig. 5-36.

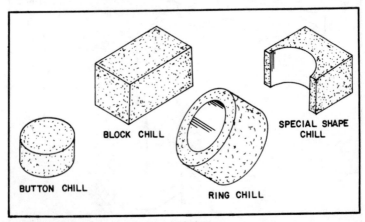

Fig. 5-39. Chills come in all sizes and shapes.

othermic compounds can be purchased so that when they are placed on a full riser, they produce heat that helps to keep the riser liquid. The risers connecting point must be large enough and not too long in order that it doesn't prematurely freeze shutting off the liquid feed metal in the riser. Hydrostatic pressure is an important factor in feeding a casting. See Fig. 5-40.

Gates and risers for manganese bronze will be well filleted with large fillets. Casting junctures shall all be well filleted. See Fig. 5-41.

CORE PRACTICE

Regular red-brass core practice is usually all you need for manganese bronze cores. Manganese bronze strength does give you some leeway in that you can get away with a stronger (more hot strength) core without getting into strains or hot tears as easily as red brass or yellow brass.

Generally, core practice does not vary over a wide range of red metals. To give various mixes for each red-metal alloy would be repetituous to say the least. I have been in shops where there were actually hundreds of various mixes (some varied in composition by

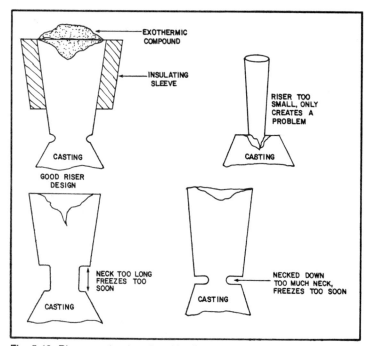

Fig. 5-40. Riser practice (good and bad).

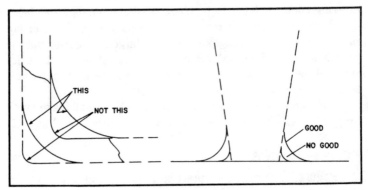

Fig. 5-41. Large fillets are a must with manganese bronze.

only a fraction of a percent of one element). This practice is extremely foolhardy and costly and adds confusion to say the least.

You might require a long, thin core, a core that will collapse easily, or a core might be surrounded by a large volume of metal subjecting it to heat for a long period. This would require a core that would not collapse too soon; it would be a core that is almost completely surrounded by metal, with a relatively small exit area.

To remove the core is still yet another problem. In this case, you would need a core that would retain its shape until the metal has set, then collapse completely so it could just about be poured from the cavity.

You might require a core with more green strength due to its configuration (to keep its shape until it is baked,etc).

If I were to give you a core mix for red-brass valves—70 parts lake sand, 30 parts silica sand, 1 part core oil, 5 percent moisture, 100 grain fineness on the sand—you'd have a core mix that should give you a green permeability of approximately 21, a baked perm of approximately 23, green compression strength approximately 4.1 PSI, baked tensile strength of 32 PSI. All this means nothing. What's missing is the design of the valve, section thickness, alloy, pouring temperature. To put it simply, the core mix is formulated to fit the casting; it is not simply picked from a list of core mixes. The various mixes given in texts (mine included) are simply guidelines. It's up to you to vary them if necessary to fit the job.

The furan no-bake process is easy to use over a wide range of casting design and alloys. You have a binder, a catalyst (often called an activator) and sand. If you require more permeability, use a coarser sand. If you need more collapsibility, use less binder. If you need more dry strength, use more binder. If you need a refractory core to prevent burn in or if you need a very smooth finish, use a good refractory core (wash) coating.

You have a large range of ways and processes available:

- [ ] Shell cores.
- [ ] Oil cores.
- [ ] Furan cores.
- [ ] Oil no-bake.
- [ ] Hot box process.
- [ ] Oil oxygen process.
- [ ] Silicate-bonded cores (CO_2 Process).

It can become very confusing. Keep it simple; furan and oil cores will meet your requirements.

Core washes also run all over the place, and there are hundreds of formulas and ingredients. Each foundry supply house has its pets. All claim theirs is the best and most reliable. You have carbonous washes that are graphite, pulverized hard coal, pulverized soft coal, coke, coal tar, carbon black, and petroleum pitches alone or in various combinations.

The so-called noncarbonous washes are silica flour, various clays, talcs, fly ash, zircon flour, magnesite, silicates, aluminum oxides, and asbestos. The bonds are water based, oil based, and solvent based.

The water-based bonds are organic bonds such as starches, dextrin, sugar, and molasses. The water-based washes consist of water, a binder, and a refractory material. A simple example would be 1 part molasses and 10 parts water, to which you would add graphite to the desired consistency degrees Baumé. This would depend on how thick you want the wash or how you intend to coat the core (dipping, spraying or brushing).

Baumé scale is a scale of specific gravity (SG) of liquids solutions and mixes.

$$\text{Degrees Baumé} = \frac{144.3 \ (SG-1)}{SG}$$

Degrees Baumé is read off of a hydrometer designed for that purpose and calibrated in degrees Baumé.

Cores are coated with a water-based wash. The wash must be dried to remove the water and set the binder. This cements the refractory coating to the core surface. This can be accomplished by spraying the core when the core is removed from the oven and still hot. The residual heat in the core will dry the wash and set the binder. If the core is too hot when sprayed, steam will cause spalding of the wash. If too cold, it will not completely dry.

Green cores can be sprayed and the wash is then dried during the baking cycle of the core. The cold cores can be dipped, sprayed or brushed and returned to the core oven and oven dried, or they can be dried with a soft torch. This can be tricky.

The inorganic binders (not water soluble) include clays, bentonites, and oxychloride. Oxychloride, also called sorel cement, is a composition of magnesium chloride, $MgCl_2$ and calcined magnesia.

The preceding binders are used with a carbonous refractory such as graphite or a noncarbonous refractory such as talc and the vehicle is isopropyl alcohol. The wash is added to the isopropyl alcohol and mixed to the desired Baumé, applied to the core, and then ignited with a match or flint lighter. The alcohol burns off, and the heat produced by this ignition sets the binder in the wash. Isopropyl alcohol is used in place of ethanol because it is not expensive. With most no-bakes and furans, you use this type of wash in place of a water-base wash. My favorite is called Zirc-O-Graph (A). It is purchased in a paste form for easy mixing. As the name indicates, a mixture of zirconium and graphite is for the refractory coating material (the binder?) and the A indicates the vehicle used is alcohol. This wash is the closest to what I have found that you could call a universal wash. I have used it on both ferrous and nonferrous work over a very broad weight range and section thicknesses. I also have used it as a green or dry sand mold wash with equal success.

Generally I will mix it to 20 to 25 degrees Baumé for spraying, 25 to 36 degrees Baumé for dipping and 36 to 50 degrees Baumé for brushing. You have to fiddle around with the Baumé to suit yourself.

Warning: Alcohol is explosive and very flammable to say the least. Work in a well-ventilated area away from any source of ignition, pilot lights, or cigarettes. Apply the wash and ignite. When the alcohol has burned off, don't pick up the core too soon. It gets quite hot.

I use this wash regardless of the core mix (oil, sand, oven-cured core, or no-bake). The point I am trying to make here is that you do not need a large collection of various core mix formulas or core washes to cover one very large range and type of castings (also alloys). Sam Pitre and I operated a nonferrous job shop for years.

We cast a wide range of work in everything from yellow brass to aluminum bronze, from a few ounces in weight to work that ran one ton and over. We had three grades of core sand, and linseed oil was used as a binder. We had some bentonite and corn cereal for green strength. For a core wash, we used molasses water and graphite or silica flour. We had three grades of molding sand (green sand) that we got from the banks of the Red River up around Alexandria, Louisiana for $2 a dump-truck load.

Don't complicate things. If I were to start up a small jobbing brass foundry today, I would buy 70 mesh washed and dried silica sand for cores, use a furan no-bake binder, and 90 to 120 mesh washed and dried silica bonded with cedar heights clay about 12 percent—and simply go to it.

MELTING

Unless you are doing very large work requiring large amounts of manganese bronze for a single casting, you are much better off with an individually fired crucible furnace or several such furnaces. Open flame furnaces where you have such a high zinc content require great care and experience. The zinc loss could be a real problem. I don't mean by this that you cannot melt economically in an open-flame furnace. For large work this is the way to go. Don't try it without getting someone who has the know-how to run a few heats with you.

Two-up pouring is quite common with medium and larger manganese bronze castings. This allows you to fill the mold with a large volume of metal in a minimum time with a minimum of turbulence. See Fig. 5-42.

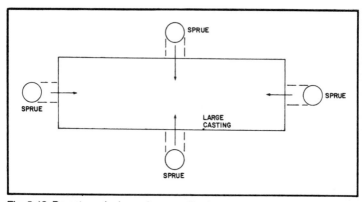

Fig. 5-42. Do not overlook pouring a casting from two or more separate gates.

The biggest controversy involving the melting of manganese bronze is the practice of flaring. Flaring is the practice of bringing the metal up to the temperature where the zinc in the alloy is actually boiling. This condition is evident when pips of flame and zinc oxide are coming off of the surface of the melt. See Fig. 5-43.

The to-flare-or-not-to-flare argument has been around—I would imagine—from crucible #1 of manganese bronze and it is still going strong. Regardless of various theories for and against flaring, you must—when melting manganese bronze—maintain careful control of the chemical composition. This is especially important with copper and zinc content. It stands to reason, as you flare off the zinc, that you have a zinc loss that will change the zinc to copper ratio of a given alloy.

This practice of flaring or heating the manganese bronze until the zinc distills off is of questionable value. The lost zinc must be replaced or you no longer have the alloy you started with nor would you have the same physical properties. If you do flare, how much

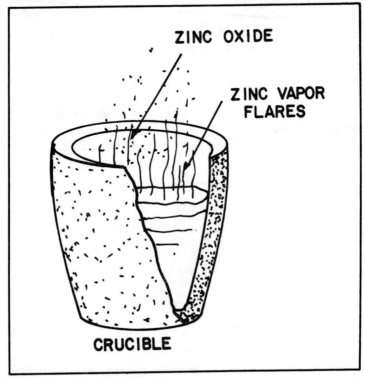

Fig. 5-43. Flaring manganese bronze.

zinc is lost and must be replaced? I have seen various figures on this; none of them makes much sense. The most prevailing figure is that 1.5 pounds of zinc is lost per 100 pounds of metal during flaring.

The fallacy of this figure of 1.5 pounds zinc loss is how hot was the metal superheated above pouring temperature and for how long? How long did the flaring take place? How much zinc will distill off per minute? Based on the surface area of the flaring melt, the only way you could tell how much zinc was lost for a given heat would be to know exactly what the zinc was prior to melting the charge, and then do a quantitative zinc assay on the casting or gates.

This would be more or less postmortem unless you had the on-the-spot equipment to make this assay prior to pouring. Then you could adjust the melt by the required zinc addition. This equipment is available but extremely expensive—and useless—unless you are a huge operation pouring many tons per day. The cost is prohibitive.

Back to flaring, there are times when pouring light work to very light work requires bringing the metal to a temperature above its incipient flaring point. Avoid this wherever possible.

The moment the metal starts to flare the best practice is to remove it from the furnace. Skim it; a thin film will form over the metal. When it cools to the point that, when you skim it very gently only a slight flaring is noted, it is ready to pour. Pour with the crucible or ladle as close to the sprue as possible and choke.

The point of incipient flaring is about 1850 degrees for low-tensile alloys and 1950 degrees for high-tensile alloys.

Fluxing and Deoxidizing is not required or practiced with manganese bronze.

IMPURITIES

Lead is the chief impurity in manganese bronze heats. Lead is excess of 0.20 percent will adversely affect the physical and mechanical properties of the metal.

If the casting is not going to be used where it will be subjected to severe stress, you can go as high as 0.40 percent lead as an impurity. The best bet, however, is to keep it below 0.20 percent. Many operators will use the same crucible to melt yellow brass, red brass and then manganese bronze (also the other way). I agree crucibles are expensive, but you should never cross melt. Have a crucible for each alloy and use it only for that alloy. The same goes for ladles. You can get away with it if the alloys are compatible, but even here you take a chance. Contaminating a heat is easy to

117

do with ladles and crucibles, and even more so with rotary furnaces.

Poor housekeeping will also cause problems. Gates, risers, and scrap must be carefully segregated. All it takes is a red brass riser or gate tossed into the manganese bronze bin. Then it is too late.

Because manganese bronze is one tough, strong metal, the removal of the gates and risers becomes a real problem that can be costly. If you do not account for this operation and the time and cost when figuring the job, you can come up a loser even though the casting is excellent.

A plus is that manganese bronze does not penetrate the sand. All you have to do is wire brush the casting. You cannot remove the risers with a torch. Bandsaw blades are available with the correct tooth design for degating manganese bronze and for use as abrasive cutoff wheels.

Due to the tendency to dross, surfaces that require machining should have enough meat in the event that you have some surface dross inclusions. Allowance for machining should not be less than 1/4 of an inch.

Manganese bronze is not what you would call hot short. It can be bent about double at room temperature or at a bright red heat 1500° without fracturing. At 1000°F, you have a crucial point where the metal will break like a soda cracker. The fracture of manganese bronze is a very fine-grained, buff color.

The fineness of the grain is dependent upon the cooling rate. The faster the cooling rate the smaller the crystallization structure.

When yellow spots show mixed with the buff color, this indicates excessive zinc in the alloy and/or lack of provision for liquid shrinkage (risers ineffective). If you heat a piece of manganese bronze to the crucial heat of 1000°F, break it and drop it into water (at this heat). Chilling is rapid and the fracture will then be a bright lemon yellow.

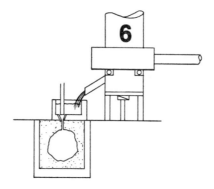

Tin and Leaded-Tin Bronzes

Tin bronzes are called the only true bronzes. These bronzes are widely used for pressure work where the casting is subject to liquid or gas pressures (valves, pumps,etc.). Tin and zinc added to copper increases its strength and hardness. Phosphorus is also a hardener in minor amounts and lead will improve the machinability and wearing properties.

I am going to cover the six most-used alloys, and you should note the different properties against their composition differences. The most common or widely used of the six is called gunmetal, and it is a lead-free alloy (88 percent Cu, 10 percent Sn, 2 percent Zn). The most common name for this alloy in the foundry is eighty-eight, ten, and two. To limit your stock of various alloys in a small foundry, if you need a tin bronze for a pressure job now and then, this is the alloy to carry.

Your gunmetal, 88 percent Cu, 10 percent Sn, 2 percent Zn and the alloy called G bronze, 88 percent Cu, 8 percent Sn, 4 percent Zn are your only real tin bronzes. See Table 6-1.

FOUNDRY PRACTICE

These alloys have a wide freezing range and require risers and/or chills on medium and large work. The requirements to provide feed metal to compensate for shrinkage fall about halfway between the red brass group and the manganese bronze.

The tin bronzes are easy to cast, there are problems such as

Table 6-1. Alloys.

Name & Composition	Properties	
G Bronze 88% Cu 8% Sn 0% Pb 4% Zn	Weight lbs/in^3 Patternmaker's shrinkage Solidification range Pouring temp. light work Pouring temp. heavy work Tensile strength Yield strength Elongation % in 2" Brinell hardness (500 kg) Machinability Heat treatment	0.318 3/16" per ft. 1832-1570° F 2100-2300° F 1920-2100° F 45,000 PSI 21,000 PSI 30% 70 30 No response
Gunmetal (SAE 62) 88% Cu 10% Sn 0% Pb 2% Zn	Weight lbs/in^3 Patternmaker's shrinkage Solidification range Pouring temp. light work Pouring temp. heavy work Tensile strength Yield strength Elongation % in 2" Brinell hardness (500 kg) Machinability Heat treatment	0.315 3/16" per ft. 1830-1570° F 2100-2300° F 1920-2100° F 45,000 PSI 22,000 PSI 25% 75 30 No response
Navy "M" 88% Cu 6% Sn 1.5% Pb 4.5% Zn	Weight lbs/in^3 Patternmaker's shrinkage Solidification range Pouring temp. light work Pouring temp. heavy work Tensile strength Yield strength Elongation % in 2" Brinell hardness (500 kg) Machinability Heat treatment	0.312 3/16" per ft. 1810-1518° F 2100-2300° F 1920-2100° F 40,000 PSI 20,000 PSI 30% 65 42 No response
Leaded "G" Bronze 87% Cu 8% Sn 1% Pb 4% Zn	Weight lbs/in^3 Patternmaker's shrinkage Solidification range Pouring temp. light work Pouring temp. heavy work Tensile strength Yield strength Elongation % in 2" Brinell hardness (500 kg) Machinability Heat treatment	0.317 3/16" per ft. 1830-1570°F 2100-2300°F 1920-2100°F 40,000 PSI 20,000 PSI 25% 70 42 No response

Leaded Tin Bronze 87% Cu 10% Sn 1% Pb 2% Zn	Weight lbs/in³	0.315
	Patternmaker's shrinkage	4/16" per ft.
	Solidification range	1800-1550° F
	Pouring temp. light work	2100-2300° F
	Pouring temp. heavy work	1920-2100° F
	Tensile strength	44,000 PSI
	Yield strength	20,000 PSI
	Elongation % in 2"	30%
	Brinell hardness (500 kg)	70
	Machinability	40
	Heat treatment	No response

Leaded Tin Bronze SAE 63 88% Cu 10% Sn 2% Pb 0% Zn	Weight lbs/in³	0.317
	Patternmaker's shrinkage	3/16" per ft.
	Solidification range	1800-1550° F
	Pouring temp. light work	2100-2300° F
	Pouring temp. heavy work	1920-2100° F
	Tensile strength	42,000 PSI
	Yield strength	21,000 PSI
	Elongation % in 2"	20%
	Brinell hardness (500 kg)	77
	Machinability	40
	Heat treatment	No response

drossing to worry about. You will note that the solidification range varies from 250° to 292° (quite a wide range). This must be accounted for when designing the gating and feeding system. Main uses are for valves, marine hardware, fittings, bearings, piston rings, gears, steam fittings, bushings, pump impellors, and machine nuts. There are various other leaded tin bronzes with a lead content from 7 percent to 25 percent Pb. Because these are primarily bearing bronzes with low tensile strengths I will cover them separately.

Tin bronzes consist of a solid solution matrix. Dentritic segregation occurring during solidification promotes separation of the eutectoid at about 7 percent Sn and over. This accounts for the good bearing properties of these tin copper alloys. This condition gives you a hard precipitate structure in a softer solid solution matrix, and thus a good strong bearing surface.

Solid Solution. A solid solution is the result of the absorption (or combining) of one or more metallic elements or compounds without changing the solid phase of the alloy.

Matrix. Matrix, with reference to metallurgy, is the crystalline phase of an alloy in which the other phases are contained.

Dentritic (dentrite). A crystal formed during solidification,

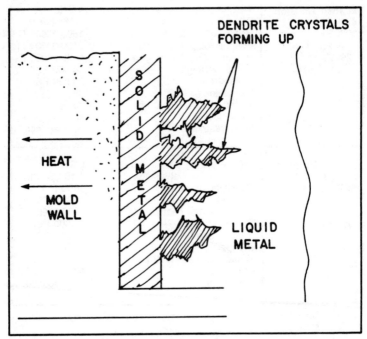

Fig. 6-1. Formation of dentrites during solidification.

dentritic has many branches of a pine tree nature; from the Greek word dendron meaning tree. See Fig. 6-1.

These alloys shown in Table 6-2 are expensive due to the high tin and copper content.

MOLDING SAND

The same sands used for red brass are suitable for the tin bronze and leaded tin-bronze alloys. In general, fine-grain sands

Table 6-2. Alloy Forms.

Ingot #225	Ingot #245	Ingot #210
88% Cu	88% Cu	88% Cu
8% Sn	6% Sn	10% Sn
0% Pb	1.5% Pb	0% Pb
4% Zn	4.5% Zn	2% Zn
Ingot #230	Ingot #215	Ingot #206
87% Cu	87% Cu	88% Cu
8% Sn	10% Sn	10% Sn
1% Pb	1% Pb	2% Pb
4% Zn	2% Zn	0% Zn

with a permeability of 20 + have a clay content 10 percent to 15 percent with a green compressive strength of 7 to 9 PSI tempered 5 percent to 6 percent moisture. A synthetic sand of 100 to 120 mesh sharp silica bonded with 4 percent southern bentonite also will work fine.

There sometimes is confusion as to sand classification, and you will find recommended sand specifications for various alloys and casting weights. Table 6-3 shows two for tin bronze. You will find formulas calling for a sand by mesh, AFA grain size, or by the name of the area where the sand comes from. Examples are Portage GFN 40, N.J. #60, Ottawa crude, Dorchester #8M, Crescent (Michigan) sand, or simply bank sand or fine sand.

You could go crazy trying to locate and sort the various mixtures. The most confusing factor is the AFA grain size. It really doesn't mean too much because you can have two sands that have the same AFA grain fineness number and yet differ in their properties by a country mile. You can have an AFA grain fineness number sand of 140—and it will produce excellent castings that are smooth—or you could have another sand AFA (GFN) grain fineness number 140 that is like corn cobs and produces rough castings.

So where does that leave you? The AFA grain fineness number—defined by the American Foundrymen's Society as the grain fineness number of a sand—is approximately the number of meshes (screen size) per inch of the sieve which will pass the sample if its grains were of a uniform size (the average of the sizes of grains in the sample). It is *approximately* proportional to the surface area per unit weight of a sand, exclusive of clay.

The key words are *if, average,* and *approximately.* What does this tell you? Well not much. It only expresses the average grain size of a given sand. What you should know is the size distribution of the sand grains.

Let's look at two sands that have the same AFA GFN. We have a series of screens (a set) and a pan. See Fig. 6-2. We have a set

Table 6-3. Tin Bronze Sand Specifications.

Grain size AFA: 140-180 Permeability: 5-20 Green compression: 7-9 PSI Clay content: 5-20% Moisture: 5-6%	**120 mesh sharp sand:** 100 lbs. Southern Bentonite: 5.0 lbs. Sea Coal (very fine): 4 to 5 lbs. Permeability: 30 to 40 Green Compression: 6-7 PSI Moisture: 4%

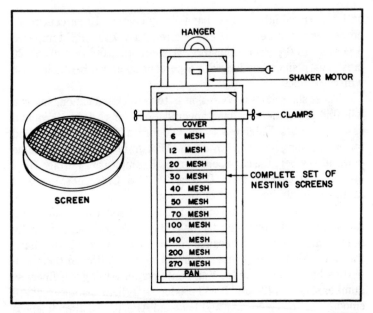

Fig. 6-2. Typical sand screening set up to classify the sand.

of 11 nesting screens (sieves) and a pan. The top screen is 6 mesh (6 openings per square inch), the next screen is 12 mesh, then 20 mesh, 30 mesh, 40 mesh, 50 mesh, 70 mesh, 100 mesh, 140 mesh, 200 mesh, 270 mesh, and a bottom pan.

If you place a 100-gram sample on the top screen and shake the screens, the various sizes of sand grains will classify themselves by passing through the screens with larger openings than the grains and they will be retained on the screen which has smaller openings than the grains. You wind up with the various different grain sizes retained on the various screens. Anything smaller than 270 mesh will wind up on the pan.

By weighing the sand retained on each screen you come up with a set of percentages. What's more important from these figures is that you can get a graphic look at the grain distribution curve. That is really the most important figure you need to know.

Let's look at two different sands; both have an AFA GFN of 60. I will spare you the mathematics of arriving at a GFN of 60 for both sands. Table 6-4 shows the grain distribution difference. It is obvious that these two sands are not the same. With #1 sand, 62 percent is retained on three adjacent screens, and on sand #2, 89 percent is retained on three adjacent screens.

Table 6-4. Grain Distribution.

Screens	100 Grams Sand #1	100 Grams Sand #2
On screen #6	0 grams	0 grams
On screen #12	0 grams	0 grams
On screen 20	0 grams	0 grams
On screen 30	1 gram	10 grams
On screen 40	24 grams	1 gram
On screen 50	22 grams	24 grams
On screen 70	16 grams	41 grams
On screen 100	17 grams	24 grams
On screen 140	14 grams	7 grams
On screen 200	4 grams	2 grams
On screen 270	1.7 grams	0 grams
On PAN	0.3 grams	1 gram

The GFN doesn't tell you much. The GFN of a sand is approximately the number of the sieve that would just pass the sample if all the grains were the same size. But they are not. Nevertheless, the percentages on each screen tell you what we want to know.

It takes the proper percentages of grain finenesses in the correct distribution to give you a good working sand that will bond

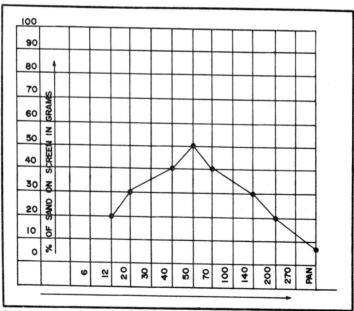

Fig. 6-3. A typical sand graph.

up properly, have sufficient permeability, rammability, and green compression strength. Let's look at it graphically (see Fig. 6-3).

The most desired grain distribution for a good working sand is one that has fairly wide grain distribution over 7 or 8 screens, with 60 perent to 70 percent of the sand retained on three adjacent screens forming a single peak. See Fig. 6-4.

Equipped with a set of standard screens, a shaker, and scales, it is no problem to quickly determine what a given sand looks like and whether or not the sand would be any good as a casting sand. This equipment also allows you to blend different sands in order to come up with what you are looking for.

With natural bonded sands, you will have to wash out the clay, dry the sample, weigh it, and then do your distribution graph. As you use a molding sand, the fines increase (as does the coarse material).

The fines come from air borne dust, blackings, crushed grains, and parting materials. Coarse material usually comes from burned core sand.

Your sand is changing all the time as you use it. You can track it with a set of sieves and make corrections to bring it back on track before it goes off too far and starts to run you into casting and molding problems. Never buy a sand without a screen analysis or you should do your own screen analysis before you buy.

When Sam Pitre and I had our shop, we would prospect for sand. When we found one with the correct grain distribution but too weak (insufficient binder), we would simply add additional clay bond, bentonite, bonding clay, or fire clay. Once you understand the molding-sand picture and what's required for what metal and temperature, things become rather simple.

Chapter 2 in *The Complete Handbook of Sand Casting*, TAB Book No. 1043, provides additional information on molding sands. Once you understand the use of sand, picking a good sand for your use is duck soup. It is not necessary to call out a sand mix for every alloy. It stands to reason that if you are pouring steel at 3000°F, the steam and gas generation is going to be not only fast but under a lot of pressure. You are going to need a much coarser sand (thus higher permeability) to get the gasses off with little resistance. The pouring temperature for steel is quite hot so the sand is going to have to be extremely refractory. Otherwise it will simply melt and adhere to the casting.

If the casting is very heavy, both in section and total weight, the sand mix must be extremely high in green strength to with-

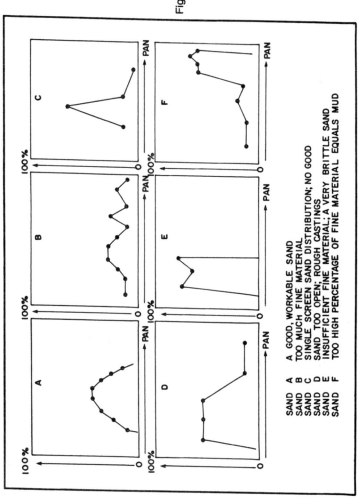

Fig. 6-4. Some typical sands graphed.

SAND A   A GOOD, WORKABLE SAND
SAND B   TOO MUCH FINE MATERIAL
SAND C   SINGLE SCREEN SAND DISTRIBUTION; NO GOOD
SAND D   SAND TOO OPEN; ROUGH CASTINGS
SAND E   INSUFFICIENT FINE MATERIAL; A VERY BRITTLE SAND
SAND F   TOO HIGH PERCENTAGE OF FINE MATERIAL EQUALS MUD

stand the tremendous forces to which it will be subjected so that the casting cavity will not be deformed. Also, the sand must have sufficient hot strength to prevent the binder (clay) from breaking down prematurely, but not be too strong in hot strength that it would restrict the movement of the casting as it shrinks during solidification (causing hot tearing).

For steel you might use a sand as coarse as 40 mesh. With aluminum, which is poured at 1200°F to 1500°F, you can pour it in extremely fine sand with low permeability and low green and dry strength. As far as the sand being refractory enough for aluminum, you probably could not find a sand of any type that would not be refractory enough to do the job.

FACING

Very seldom would you have the need to face a mold for tin bronze and leaded tin bronze (only in special cases such as extremely heavy work). You might elect to dust small, light-work molds with wheat flour in a parting bag and heavier work with a blacking bag (graphite).

GATES AND RISERS

Gates and risers usually are somewhat larger than red brasses. The same rules apply for feeding a gate into a heavy section (riser receives hot metal last to be effective).

CORE PRACTICE

As with red brass, whether to coat the cores depends upon the casting design and the surface requirements of the cored surface.

MELTING

These alloys can be melted in any suitable crucible or open-flame furnace. Melt fast and do not soak and maintain a slightly oxidizing atmosphere. Raise the metal to only about 100°F above the pouring temperature. Of course, if melted in an open-flame furnace you need a good cover slag to prevent the melt from coming in contact with the products of combustion. The pouring range of from 1950°F to 2250° will suffice for the largest percentage of the work. When it is necessary to go to 2300°F, for extremely thin work, this temperature produces excessive phosphorus that will make the metal too fluid, causing penetration of the core and mold surfaces. At 2300 °F, the zinc loss will activate and will require zinc additions to bring the alloy to specification.

FLUXING AND DEOXIDIZING

In open-flame furnaces, a cover flux of lead free glass and Razorite is a good choice. Some operators prefer a flux consisting of 3 parts lime and 1 part fluorspar or soda ash and fluorspar. I would imagine that this flux could be very detrimental to the refractories and should eat up a crucible in no time.

A system once widely used (and perhaps still used) in Canada is called the Walpole flux method. This treatment was considered the best method of getting the maximum density and pressure tightness in tin bronze and leaded tin bronze for pressure work, valves, fittings, pumps, and any casting where medium and high pressure was involved. I have never tried this method.

The flux consists of 1 part by weight of copper oxide (CuO) (Dry Copper Mill Scale), 1 part by weight clean sharp silica sand (core sand), and 1 part by weight powdered fused borax (calcined borax).

The flux was mixed and put into a tight container to prevent moisture absorption until needed. Three pounds of flux was used for each 100 pounds of metal charged or enough to give you a cover when the metal became molten of at least 1/4 of an inch thick. The flux was added at the time the metal was charged. The heat melted under slightly oxidizing conditions (as rapidly as possible).

Here is the puzzler. When the melt is up to the desired temperature, the flux is stirred into the melt, and a little dry silica is thrown on the melt to thicken the slag so it can be easily removed with a skimmer.

I can see where a cover flux containing copper oxide could insulate the melt from oxidizing products of combustion, just as any patina or oxide coat on the surface of a metal prevents further oxidation by coating the metal with an oxide coat insulating it from further oxidation. But why stir it in? Here's something for you to think about. I know of several "secret fluxes" widely advertised by some foundry suppliers. They sell for high prices but they are basically copper oxide and calcined borax. The application instructions for these fluxes do not call for stirring it into the heat.

Tin bronze and leaded tin bronzes are deoxidized with the addition of 15 percent phosphor copper, 2 to 4 ounces per 100 pounds of melt just as you do with red brasses. The only difference is that the tin bronze and tin lead bronze alloys can tolerate a higher percentage of residual phosphorus. In fact, a residual phosphorus of 0.05 to 0.30 percent is desirable.

IMPURITIES

Phosphorus. An excess of phosphorus will make the metal too fluid. It also will increase the hardness.

Silicon. Silicon will cause no problems in the lead-free alloys such as 88-8-0-4 and 88-10-0-2 or in alloys where the lead is a maximum of only 0.25. When above this figure and with the leaded tin bronzes, silicon can be extremely harmful.

Warning. Beware of a red metal alloy containing lead when it is deoxidized with Phos Cu, lithium, etc. In this condition, any silicon present (as an impurity) causes lead silicate to be formed. Silicon dioxide plus lead oxide equals lead silicate Pb S_1O_3.

Aluminum. Aluminum, even in very minor amounts such as a few hundredths of a percent in tin or tin lead bronzes, is disastrous. The result is destroyed physical properties of the castings.

Antimony. At 0.25 percent and below, antimony is not harmful.

Iron. Iron will cause random hard spots in the castings.

Sulfur. Up to 0.08 percent maximum, sulfur is not harmful.

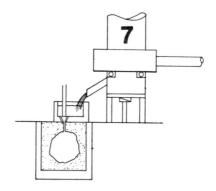

High-Leaded Tin Bronze

High-lead, tin bronzes are the bearing and bushing alloys. They run from 27,000 PSI to 35,000 PSI in tensile strength. That puts their top tensile range comparable to commercial red brass and semired brass. I have seen castings for valves, small machine parts, and the like made from high-leaded tin bronze, but this use is the exception. Where you have a shaft wear problem these bronzes can't be beat. High-leaded tin bronzes present some problems in casting due to their high lead content. Lead and copper do not form a solid solution. What you wind up with is copper crystals (dentrites) with the spaces filled with lead. If you filled in the spaces in a copper sponge with lead, you would about have it. This is how you get the good bearing properties.

Molten copper will only dissolve about 35 percent lead (liquid solution), but when alloyed with 2.5 percent nickel the solubility of lead in molten copper jumps up to almost twice this value. Because lead is for all practical purposes insoluble in the solid state when it is thrown out during solidification, the presence of some nickel and the tin gives you a fine uniform dispersion throughout the casting. This gives you the best possible high-lead tin bronze, bearing or bushing. Many high-lead, tin-bronze ingots are nickel free. Some have 1/2 percent to 1 percent nickel.

The practice used by some shops is to ring the nickel up to 1/2 percent by adding pure nickel shot. If the alloy is nickel free, some prefer to keep the high-leaded tin bronzes nickel free. For bear-

Table 7-1. High-Lead Bronzes.

Name & Composition	Properties	
Bearing Bronze SAE 660 83% Cu 7% Sn 7% Pb 3% Zn	Weight lbs/in³ Patternmaker's shrinkage Solidification range Pouring temp. light work Pouring temp. heavy work Tensile strength Yield strength Elongation % in 2" Brinell hardness (500 kg) Machinability Heat treatment	0.322 7/32" per ft. 1790-1570°F 2000-2250°F 1900-2050°F 35,000 PSI 18,000 PSI 20% 65 70 No response
High-leaded Tin Bronze SAE 84% Cu 8% Sn 8% Pb 0% Zn	Weight in lbs/in³ Patternmaker's shrinkage Solidification range Pouring temp. light work Pouring temp. heavy work Tensile strength Yield strength Elongation % in 2" Brinell hardness (500 kg) Machinability Heat treatment	0.322 3/16" per ft. 1750-1570°F 2000-2250°F 1850-2100°F 32,000 PSI 16,000 PSI 20% 60 70 No response
High-leaded Tin Bronze SAE 66 85% Cu 5% Sn 9% Pb 1% Zn	Weight in lbs/in³ Patternmaker's shrinkage Solidification range Pouring temp. heavy work Pouring temp. heavy work Tensile strength Yield strength Elongation % in 2" Brinell hardness (500 kg) Machinability Heat treatment	0.320 3/16" per ft. 1830-1570°F 2050-2300°F 1950-2100°F 32,000 PSI 16,000 PSI 20% 60 70 No response

Bushing & Bearing Bronze
80% Cu
10% Sn
10% Pb
0% Zn

Weight in lbs/in³	0.32
Patternmaker's shrinkage	1/8" per ft.
Solidification range	1705-1403°F
Pouring temp. light work	2000-2250°F
Pouring temp. heavy work	1850-2100°F
Tensile strength	35,000 PSI
Yield strength	18,000 PSI
Elongation % in 2"	20%
Brinell hardness (500 kg)	60
Machinability	80
Heat treatment	No response

Note: This alloy is probably the one most used. As I recall in my years at the game, 95 percent of the time when a leaded bronze bearing casting was called for, it was 80-10-10.

Anti-Acid Metal
78% Cu
7% Sn
15% Pb
0% Zn

Weight lbs/in³	0.334
Patternmaker's shrinkage	1/8" per ft.
Solidification range	1730-1570°F
Pouring temp. light work	2000-2250°F
Pouring temp. heavy work	1900-2100°F
Tensile strength	30,000 PSI
Yield strength	16,000 PSI
Elongation % in 2"	18%
Brinell hardness (500 kg)	55
Machinability	80
Heat treatment	No response

High-Leaded Tin Bronze
70% Cu
5% Sn
25% Pb
0% Zn

Weight lbs/in³	0.336
Patternmaker's shrinkage	1/8" per ft.
Solidification range	1700-1570°F
Pouring temp. light work	2000-2200°F
Pouring temp. heavy work	1850-2000°F
Tensile strength	27,000 PSI
Yield strength	13,000 PSI
Elongation % in 2"	10%
Brinell hardness (500 kg)	48
Machinability	80
Heat treatment	No response

ings that are going to be subjected to extreme pressure and operating conditions, you could use one of the nickel alloyed bronzes such as Ni-Vee bronze (copper 80 percent, nickel 5 percent, tin 5 percent, lead 10 percent, zinc 1 percent maximum, phos. 0.02 percent maximum). Here you have extremely strong bearing material 40,000 PSI tensile, 30,00 PSI yield, 5 percent elongation in 2 inches and a Brinell hardness of 110 and a 10 MM ball at 1000 kg load.

Table 7-1 shows six typical high-leaded bronzes. These are the basic six leaded-tin, bronze-bearing alloys. You will note as the lead percentage goes up the machinability goes up, the tensile strength goes down, the hardness goes down, with an increase of lead the solidification spread shortens. At 3 percent Pb you have a 220-degree spread. At 25 percent Pb, this spread is only 130 degrees.

FOUNDRY PRACTICE

Let's talk about these high lead tin bronzes and their solidification range of from 220 degrees for 83-7-7-3 alloy to 130 degrees for 70-5-25-0 alloy. With the top starting temperature of solidification of 1830 °F to a bottom finish temperature of 1403 °F, all this can be very misleading. One fact often overlooked by the foundryman when working with high-lead alloys is that the casting is not truly solidified (completely solid) until it is below 621 °F (the freezing point of lead). When you reach the bottom end of the solidification range of any of these six alloys, 1403 °F or 1570 °F, you have solidified Cu and Sn network filled with liquid lead droplets. Then you have a condition where the apparent shrinkage of the casting is slight. This is because it is intercrystalline shrinkage spread out over a large area rather than as a single void in the casting.

Look at it this way. If you pour a test cock out of 84-8-8-0 high-lead tin bronze that has a shrinkage of 3/16 of an inch per foot and a solidification range of 1750 to 1570 °F, and a test cock out of say straight silicon bronze 91 percent Cu, 4 percent Zn, and 5 percent Si with a shrinkage of 3/16 of an inch per foot and a solidification range of 1720 to 1540 °F, both of these alloys have the same shrinkage per foot of 3/16 of an inch and 180-degree solidification spread. You would imagine that these different alloys, in view of the same shrinkage and solidification characteristics, would act the same way as to shrinkage. But not so, see Fig. 7-1.

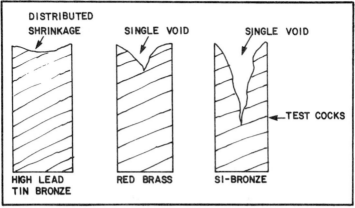

Fig. 7-1. Distributed shrinkage of high-lead tin bronze compared with red brass and silicon bronze.

The shrinkage in the top of the Si bronze and red brass test cock (Fig. 7-1) is quite pronounced. With the high leaded tin bronze it is hardly noticeable. The two alloys actually took the same shrinkage, one as a single type shrinkage and the other an inter-crystalline shrinkage. This property, that of spreading its shrinkage over the entire casting, makes high-lead, tin bronzes easy to gate. Only the very heavy castings require any risers. Remember the key words here are *apparent shrinkage* as opposed to actual shrinkage.

How about the lead's solidification point of 621 °F? All you have to do is shake out a high-lead, tin bronze casting too soon and it will bleed (sweat) lead all over the place. You will wind up with a sponge for a casting setting in a pool of lead. Many a casting in high-lead tin bronze has been reduced to scrap due to some eager beaver shaking out much too soon.

Table 7-2. Alloy Forms.

Ingot #315	Ingot #311	Ingot #326
83% Cu	84% Cu	85% Cu
7% Sn	8% Sn	5% Sn
7% Pb	8% Pb	9% Pb
3% Zn	0% Zn	1% Zn
Ingot #305	**Ingot #319**	**Ingot #322**
80% Cu	78% Cu	70% Cu
10% Sn	7% Sn	5% Sn
10% Pb	15% Pb	25% Pb
0% Zn	0% Zn	0% Zn

ALLOY FORM

Good clean scrap, if properly identified, or ingots from a smelter with heat analysis figures make up the alloys shown in Table 7-2.

MOLDING SAND

With high-lead alloys you have a real problem with metal penetrating both the mold walls and the core. The use of molding sands other than very fine grain sizes with a good distribution will accentuate the problem. Your best bet is a natural bonded sand such as a French or Windsor Locks sand, or a sand that compares closely to Windsor Locks or French. Both of these sands are fine flourlike sand that, when properly tempered, have a silky feel with no gritty feel whatsoever.

Windsor Locks sand has an average fineness of 270, permeability 10 to 18, clay content 10 percent to 20 percent, and green strength 6 to 8 PSI at 6.5 percent to 7 percent moisture. French sand is an imported sand from France, and it is quite expensive. It is available from some foundry supply houses. French sand average fineness (clay left in) is 135 to 170, average fineness (clay out) is 140 to 175, clay substance (average) is 16 percent, permeability is 18 at 7.2 moisture with green strength 7.9 PSI.

I don't expect you to run out and buy French sand or Windsor Locks sand. I do know you can find loads of river sand that will come close enough even if you have to add some bonding clay to bring up the green strength. Let's set a limit as to your specifications: AFA fineness 150 to 230, permeability at 6 percent moisture 10 to 25, clay content 10 to 16 percent, green strength 5 to 9 PSI.

I don't mean by this that you cannot get away with a more open (coarser) sand with a high permeability. Nevertheless, casting finish will be a problem.

There are various clay-carrying and clay-free fine natural sands that when bonded with clay or bentonite will fill the bill. A 120-mesh silica bonded with clay or bentonite and a few percent of very fine sea coal will produce excellent results. The sea coal will greatly reduce the tendency of metal penetration.

FACING

It is common practice to face when casting high-lead tin bronzes in order to prevent penetration. Facing the pattern with 1/2 inch to 1 inch of very fine molding sand and backing with your system sand is often practiced or a special facing sand can be made up. A good facing would be 10 parts of very fine molding sand—such as #00 albany or equivalent—1 part of goulac or dextrin, and 3 parts of airfloat sea coal. You could, if you chose, delete the goulac or dextrin and temper the dry sand and sea coal with molasses water (1 part molasses to 10 parts water) and lightly skin dry the mold.

Another choice you have is to dust the mold with a high grade plumbago (graphite) and brush it down with a soft camel-hair brush. Or you could spray black the mold with a commercial wash such as Zirc-O-Graph A or a homebrewed wash of plumbago and molasses water. If you used Zirc-O-Graph A, to set the binder or torch dry your molasses water plumbago wash to set the molasses. *Note:* foundry molasses is a cheap molasses that is usually denatured (poisoned) to keep the ants and the molders away.

GATES AND RISERS

High-leaded alloys are the simplest to gate mainly due to the intercrystalline shrinkage. In general, only the larger castings require risers to feed the casting. Because these alloys have a high specific gravity and weigh from .320 pounds per cubic inch to .336 pounds per cubic inch and are extremely liquid, they have a tendency to cut and wash. For this reason, it is often necessary to construct the sprue system and ingates in core sand or no-bake to prevent the washing of sand from the sprue or gate into the mold cavity. See Fig. 7-2.

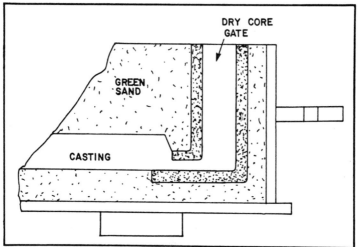

Fig. 7-2. A dry-sand core gate prevents sand washing and erosion.

A splash core or tile is often placed directly below the sprue to prevent washing (eroding the sand away). In the case of an extremely deep gating, you should step the sprue with a tile or core at the bottom of each step. See Fig. 7-3.

Foundry nails are used at the bottom of sprues, gate and, in some cases, part way into the cavity to prevent washing. See Fig. 7-4. In some work an inverted horn gate will prevent erosion (wash). See Fig. 7-5.

Gates should always be examined for evidence of washing as well as the casting, cuts and washes (the results of the sand of the mold or core eroding away and washing into the casting). The washing can come from the gates or other sections of the casting. They appear as rough spots and excess metal at the spot where the sand washed. If the gate washes, but not the casting the sand that was eroded away goes into the casting and might be visible on the casting surface. It can be trapped in the casting and show

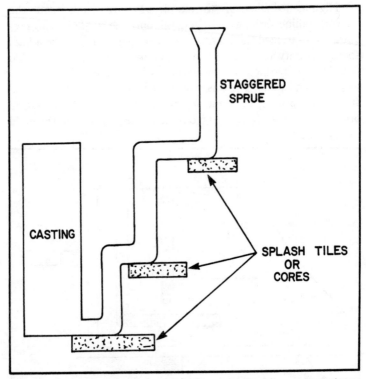

Fig. 7-3. Staggered sprue and splash cores prevent erosion and turbulence in gating a deep mold.

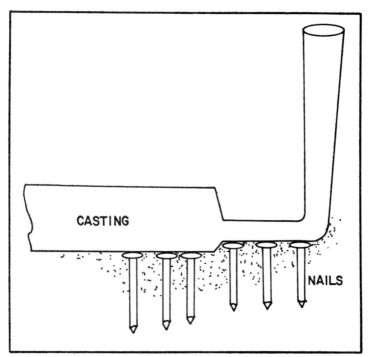

Fig. 7-4. Foundry nails prevent erosion by incoming metal flow.

up during machining or as a leaker if the casting's ultimate use is a pressure proposition a valve, a fitting, etc.

If you have a gate or sprue with excess metal, usually in the form of raised streaks, you have a wash. See Fig. 7-6. Wash can be caused by many things or conditions.

Design. If the design requires excessive metal flow over a giv-

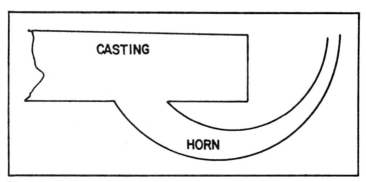

Fig. 7-5. Inverted horn gate revisited.

139

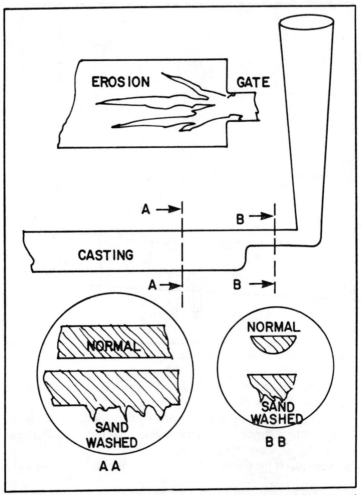

Fig. 7-6. A cross section of the defect caused by sand erosion of the gate and mold surface.

en point for too long of a duration, it might be impossible to gate without causing some spraying or nozzling of the incoming metal from the gates (try a horn here).

Gating and Risering. With excessive metal flow through the gate, you can increase the number of gates as a remedy.

Insufficient Choke. When you close down the choke, it is effective for a remedy. Where a gate directs the metal against a core or mold projection, the remedy is to move gate. A nozzling gate (pinched down at entry point) is shown in Fig. 7-7.

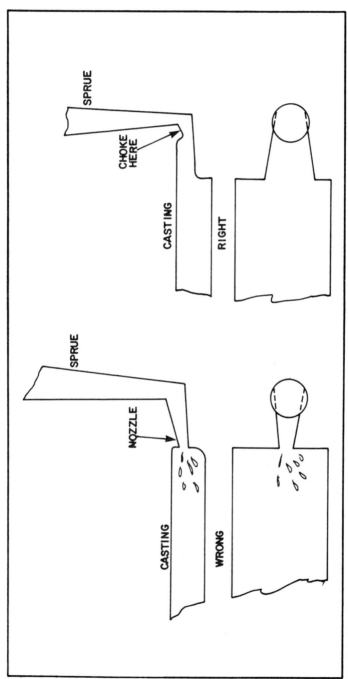

Fig. 7-7. Nozzling is a prime cause of defective castings.

141

Sand. Too much moisture boils and kicks away sand. The remedy is lower temper moisture. With sand too weak in green strength or hot strength check bond, add moisture and conditioning of your sand to correct. If permeability is too low, the remedy is to open up sand with coarser sand.

When sand contains excessive fines, check the sieves for high percent of pan material. You might be able to reduce the percent of fines by dilution. If the percentages are too high, you have to delegate it to the dump. And on and on.

That these alloys are extremely fluid, making them seeking in nature, is the big problem. If everything is not just right, this extremely fluid liquid will easily get into weak areas and dislodge the sand, core wash, and erode away the surface. Then you have this eroded material in the casting. As with any casting, you are looking for directional solidification. If risers are necessary, they must be of sufficient size that they will promote directional solidification and not cause the problem of solidifying ahead of the casting or be connected in such a manner that the connection freezes (canceling the reason and performance of the riser in the first place, which I have been harping on all along). See Fig. 7-8.

If the mold is such that the riser needed cannot be placed close enough to the casting for some physical arrangement of the casting design or rigging, you can put a shrink bob in the gate connecting the riser to the casting to keep the connection open and fluid. This bob should always be in the cope. The metal shrinks from the cope down in feeding. If you put the heat reservoir in the drag, you defeat the purpose and it will freeze off. See Fig. 7-9.

Making a hot spot at the bottom of a sprue or riser can cause problems. Why some molders do this is a mystery. It seems to be simply a habit or lack of knowledge of what's going on during solidification. See Fig. 7-10.

Some think this is a dirt catcher; not so. A blind riser is effective, but should be vented to the outside to give it some atmospheric pressure to help it feed. If it is blind, you can put a core in the top that will generate some gas pressure to help with the feed. Blind risers are available from supply houses that are made of Styrofoam and simply vaporize. This gas helps with the feeding.

In molding, the Styrofoam riser is placed on a pin on the pattern and the cope is rammed in the usual manner. When the cope is removed (drawn) or the pattern is drawn, the Styrofoam riser remains in place to do its job. See Fig. 7-11.

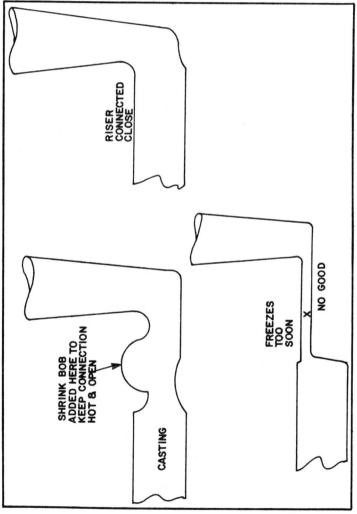

Fig. 7-8. Good and bad practice in side riser connection.

143

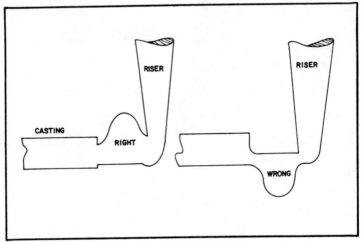

Fig. 7-9. More side riser design (good and bad).

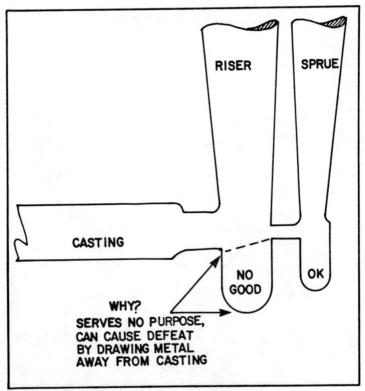

Fig. 7-10. Widely practiced but not recommended techniques.

144

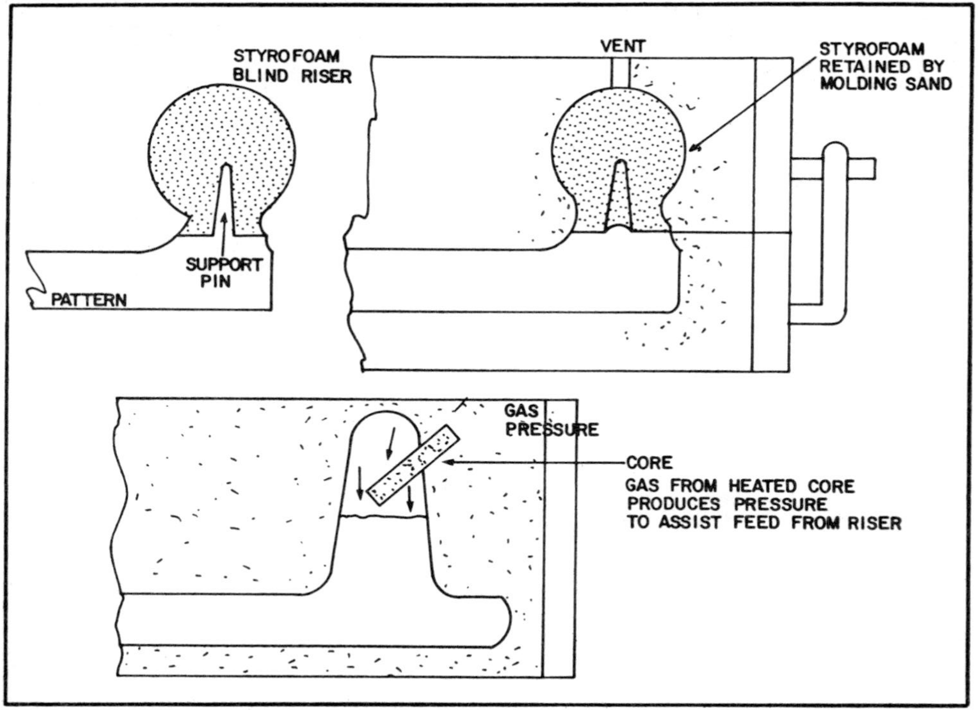

Fig. 7-11. Preformed blind risers made of Styrofoam are a useful practice.

STYROFOAM RETAINED BY MOLDING SAND

VENT

GAS PRESSURE

CORE

GAS FROM HEATED CORE PRODUCES PRESSURE TO ASSIST FEED FROM RISER

STYROFOAM BLIND RISER

SUPPORT PIN

PATTERN

Some operators cover the risers with crushed, dry charcoal to insulate the risers and keep them fluid longer.

CORE PRACTICE

Because high-lead tin bronzes are very hot short at a red heat, the cores must be fairly weak in hot strength so as to collapse soon enough to prevent hot tearing of the casting as it shrinks. That would happen if the movement of the casting, as it shrinks, were restricted by too great of a force. This also applies to the molding sand if it is too high in hot strength.

With some wood flour in the core mix, your molding sand mix will increase its collapsability. The wood flour burns, producing voids, and allows the sand to be moved by the shrinkage pressure of the casting. Cores are often hollowed out in medium and large work so that they will give away and prevent hot tearing. See Fig. 7-12.

Use low binder ratios to sand. Because cores for high-lead work require high permeability with very ample vents to prevent core gas from backing up into the liquid metal—causing gas porosity in the solidified casting—you are looking at a fairly open-coarse, grain-core structure. Now couple this type of core with the high fluidity of these alloys. You will get into metal penetration and possibly washing.

It is imperative that the cores be coated with a good core wash (preferably a good grade of plumbago wash). Dry sand or no-bake molds should also be washed with a good blacking wash. The Zirc-O-Graph A wash I have mentioned works very well with these alloys for cores and molds (dry molds or green molds).

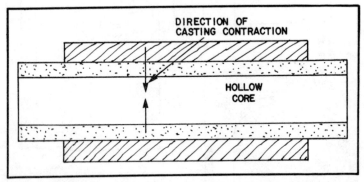

Fig. 7-12. A hollow core is good insurance against hot tearing when you are casting heavy work.

MELTING

Melt fast and cover the crucible with a cover or flux (slag) cover of lead-free glass and charcoal, a borax glass flux or borax, or glass soda ash flux. Some say no to charcoal. You can melt in open-flame or crucible furnaces with an open flame. You will need a good flux cover to prevent contamination of the melt from the products of combustion. Adjust the flame slightly to oxidizing.

FLUXING AND DEOXIDIZING

Flux as covered in melting. Use 15 percent phos. copper shot to deoxidize high-lead tin bronzes. Liberal doses of phos. copper are quite common with these alloys as high as 6 to 8 ounces of 15 percent phos. copper per 100 pounds of metal in most cases. Or phosphor-tin can be used.

The preference is for phos. copper because metal deoxidized with it, in comparison to phos. tin, seems to lessen the formation of dross during pouring. The procedure is the same as with red brass. With the alloys that contain between 15 percent to 25 percent lead, you can run into the problem where even large doses of phos. copper fail to completely deoxidize the melt. It has been found that the addition of copper oxide or lead oxide in the amount of not over 0.10 percent, five minutes before the melt is up to temperature, will help. Then add the phos. copper as usual.

Lead bleed or lead sweat and lead segregation is always a problem with these alloys, and it is accentuated very much once you get to 15 percent lead content and higher. I have talked about shaking out the casting too soon while the lead is still liquid. Nevertheless, a large casting will, in many cases, have segregation of lead into the lower parts of the casting via gravity.

I would imagine that you could have some segregation even when the metal is in the molten stage. Let's look at the atomic weights and specific gravities of our three main constituents: lead, atomic weight 207.2, specific gravity 11.35; copper, atomic weight 63.54, specific gravity 8.96; tin, atomic weight 118.69, specific gravity 7.29.

Another gravity segregation problem with high-lead tin bronzes is that the first crystals to form might sink to the bottom of the melt or casting. Impurities in the melt can also cause another type of segregation, called dendritic segregation. The rejection of impurities to the grain boundaries makes a loose segregated open structure. See Fig. 7-13.

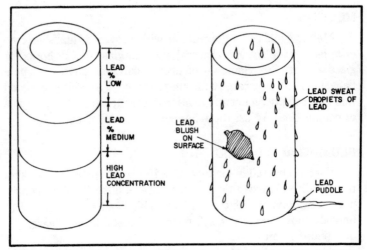

Fig. 7-13. Lead sweating and segregation is a problem with chunky, high-lead red metals.

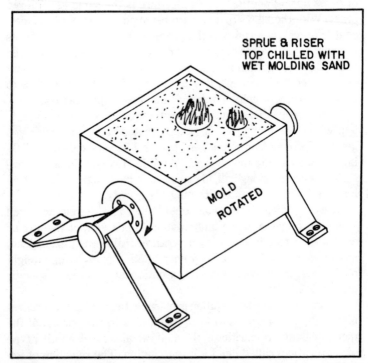

Fig. 7-14. Mold rotation during solidification is sometimes used to prevent lead segregation. This is a special case.

148

Extremely large high-lead alloys are often poured (using trunions) in steel flasks. When the copper/tin phase has solidified, the tops of the sprues and risers are frozen off with wet molding sand, and the mold is rotated slowly until the lead has solidified to prevent gravity segregation of the lead. See Fig. 7-14.

IMPURITIES

The number one impurity is silicon. We have been over the effects and chemical reaction of silicon and lead. They are simply not compatible.

In the correct amounts, phosphorus is beneficial as a deoxidizer and densifier. In excess, it can become an impurity and cause excessive shrinkage. The metal could become so fluid as to cause sand washing or fusing with the sand.

Aluminum is about as detrimental to high-lead alloys as silicon. Because aluminum causes drossing and increased lead sweating, even traces of aluminum should be avoided.

Iron will cause hard spots and slow up in machining. It should be avoided as much as possible, but minor amounts do not seem to cause problems.

Sulfur will cause pinhole porosity and general gassy castings. The origin of sulfur in a melt is usually traced to high-sulfur fuels when melting with cheap oil or coke. Check the sulfur content of your oil or solid fuel before you buy it.

Stick to good, basic practice with these alloys. Use generous fillets and avoid taking on a casting that is obviously a poor design not suited for these alloys. You will come in contact with some designs that are not suitable to cast in any metal or metal alloy. Believe me I have seen many.

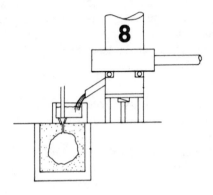

Silicon Bronze and Silicon Brass

Silicon is about the worst hazard to have in any brass or bronze that contains lead. Silicon consists of dark-colored crystals that are soluble in a mixture of nitric and hydrochloric acid, hydrofluoric acid, and alkalies. The specific gravity is 2.33 melting point 1410 °C.

One of its many uses is alloying agent for steels, aluminum, bronze, copper, and iron. Sand is silicon dioxide SiO_2 (sand, quartz, fling, and datomite). Silicon does not occur free in nature. Its derivation is by heating sand and coke in an electric furnace.

Silicon is available as hardeners (master alloys) in various forms such as shot, waffle, ingot and in various combinations and percentages to fit a particular need—such as aluminum-silicon, copper silicon, ferro filicon (iron and silicon).

Silicon hardeners are widely used both in ferrous and nonferrous applications. With cast iron, silicon can be introduced in the iron as a ferro-silicon shot or as a special high-silica pig (silvery pig). It is the element that is especially important in fixing the proportioning of free carbon and combined carbon. In the absence of silicon, when the iron solidifies the carbon will remain in chemical combination (solid solution) producing a glass hard white iron.

As you add silicon, some of the carbon will be thrown out when the iron solidifies as free graphite (called graphitic carbon), giving the characteristic gray fracture to the iron. At 3.00 percent silicon, the fracture will be dead gray. It is by this method that the iron founder controls the physical properties of the cast iron.

We have been working with alloys that consist of various percentages and combinations of metals such as Cu, Sn, Pb, Zn. Now I will discuss alloys of various metals and a nonmetal silicon. Very simply, silicon brass and silicon bronze consist of a copper-silicon solid solution with a silicide compound according to composition. The suffix IDE is used in naming compounds comprised of two elements.

There are a wide range of silicon copper alloys (compounds) with a silicon percentage from only a trace to a silicon percentage of 6 percent plus. Some alloys are nearly pure copper; it is simply fluxed with silicon used for strong, high-strength electrical wires and electrical hardware. A typical alloy of this type would be 98.55 percent Cu, 1.40 percent Sn and 0.05 percent silicon. Hard drawn wire of this combination will have a tensile strength of 92,000 PSI minimum. On the other end of the scale, you could have a casting silicon copper alloy that would look like this: 91.75 percent Cu, 6 percent Si, and 2.25 percent Fe. Tin alloy is called Cu-Si-Fe bronze.

Silicon copper combinations—with various other alloying elements such as iron, zinc, manganese, phosphorus, aluminum, nickel, tin—go by a wide variety of names such as P.M.G., Everdur Tombasil, Herculoy, Phono bronze, Phono electric alloy, Duronze 708, Silnic Bronze, White Tombasil, Vulcan Bronze, Olympic Bronze, Olympic Bronze G, Doler Brass, Jacobs Alloy, and Kuprodur. These trade names are but a few. Each company has its own trademark. Many of the alloys are very similar in both chemical composition and physical properties. In this chapter, I cover four of American Smelting and Refining Company's silicon copper alloys. By this I do not mean these silicon copper alloys are superior to the competitive silicon copper alloys out there produced by other smelters. It is not an endorsement.

Silicon copper casting alloys are made by adding silicon in the form of silicon-copper hardeners in the correct percentage to give the alloy the desired silicon percentage. Most founders purchase ingots of the desired alloy. If you have a good supply of #1 clean melting scrap copper, I don't see why you could not come out quite well by purchasing some silicon-copper master alloy and produce your own.

If the spread is wide enough it is the way to go. Large smelters are faced today with stiff worldwide competition, high operating costs brought on by high labor, OSHA, and EPA. This makes things look better for the small foundry.

Silicon brass and bronze are growing in popularity as a casting metal choice. There are many reasons for this. Silicon is an easy metal to cast, it has excellent fluidity, it is easily melted, there is no smoke or fumes, there is little to no dross formation, casting surfaces (as cast) are unusually clean, smooth and free from sand, there are low melting and pouring temperatures, and the metal takes a very high polish.

The polished metal is a rich, gold color with excellent resistance to corrosion. It can be sand cast, or you can use permanent molds (cast iron molds) and die cast, free from as porosity (caused by absorbed gasses). Usually (99 percent of the time) no deoxidizers are needed. The castings are strong with a fine-grained dense structure and ductile. Castings with great detail and sharpness are easily produced. The castings are easily worked hot or cold, welded and plated. Silicon bronzes are less expensive than tin bronzes because tin prices fluctuate wildly whereas silicon does not.

Scuptors prefer these alloys for the preceding reasons plus the castings, being ductile, can be bent without breaking or fracturing. Small protrusions such as the fingers of a statue can be bent to change position or returned to their original position should they accidently get bent due to a fall or blow.

Silicon bronze art castings can be easily chased (finished) and patinaed. If you have a small, backyard foundry or a foundry in your garage, the lack of smoke or fumes when melting silicon bronze is a large plus when it comes to not attracting your neighbors, the fire department, or OSHA. You crank up a batch of yellow brass or manganese bronze and the zinc oxide (dense white smoke) bellowing from you stack or window will bring them running.

Another plus is that you have a very low melt loss with melting and remelting. Also silicon copper alloys can be remelted castings and gates and risers with little or no change in physical properties. The use of these alloys are legion: bearings, art bronzes, architectural work, machine parts, gears, rocker arms, valves, bells, propellers etc. Silicon bronze bells ring with a very clear tone and have a great carry. Ring carry is the length of time the bell lasts in time after being struck.

Four typical silicon copper alloys are shown in Table 8-1. Silicon brass is defined as a copper-base alloy containing 0.5 percent Si and over 3 percent Zn. Silicon bronze is defined as any copper alloy that contains Si with a copper content not to exceed 98 percent.

Table 8-1. Silicon Copper Alloys.

Name & Composition	Properties	
Herculoy Regular(TM) 92% Cu 4% Zn 4% Si	Weight lbs/in³ Patternmaker's shrinkage Solidification range Pouring temp. light work Pouring temp. heavy work Tensile strength Yield strength Elongation % in 2″ Brinell hardness (500 kg) Machinability Heat treatment Note the low machinability compared with red brass	0.302 3/16″ per ft. 1780-1580°F 2050-2250°F 1900-2050°F 55,000 PSI 25,000 PSI 30% 85 40 No response
Herculoy Gear Bronze(TM) 91% Cu 4% Zn 5% Si	Weight lbs/in³ Patternmaker's shrinkage Solidification range Pouring temp. light work Pouring temp. heavy work Tensile strength Yield strength Elongation % in 2″ Brinell hardness (500 kg) Machinability Heat treatment Note the jump in hardness with 1% additional silicon.	0.30 3/16″ per ft. 1720-1540°F 2000-2250°F 1850-2050°F 65,000 PSI 30,000 PSI 25% 125 30 No response
Everdur 1000(TM) 95% Cu 4% Si 1% Mn	Weight lbs/in³ Patternmaker's shrinkage Solidification range Pouring temp. light work Pouring temp. heavy work Tensile strength Yield strength Elongation % in 2″ Brinell hardness (500 kg) Machinability Heat treatment	0.295 3/16″ per ft. 1850-1700°F 2100-2250°F 1900-2050°F 50,000 PSI 20,000 PSI 30% 80 35 No response
Tombasil(TM) 82% Cu 14% Zn 4% Si	Weight lbs/in³ Patternmaker's shrinkage Solidification range Pouring temp. light work Pouring temp. heavy work Tensile strength Yield strength Elongation % in 2″ Brinell hardness (500 kg) Machinability Heat treatment	0.299 3/16″ per ft. 1680-1510°F 1900-2100°F 1750-1900° F 67,000 PSI 30,000 PSI 21% 115 50 No response

Note: Silicon bronzes are far from new alloys. How far back they go I don't know. I checked back as far as 1927, and keep finding them, so I stopped looking. The 1927 Silicon Bronze had two basic alloys: #1 90% Cu, 9% Sn, 1% Si. #2 98% Cu, 1.40% Sn, 0.60% Si.

This is how the ASTM classifies it (which really leaves you somewhat in doubt as exactly what is what).

FOUNDRY PRACTICE

Foundry practice is very much like that practiced with red brass. You will note the patternmaker's shrinkage for silicon bronze and silicon brass is the same as for red brass (3/16 inch per foot). But look at the difference in the solidification range. Red brass 85,5,5,5 has a 280-degree spread, commercial red has a spread of 290 degrees, Herculoy has a 200-degree spread, and Everdur has only a 150-degree spread. If you look at tin bronzes and manganese bronzes, you will note that with silicon bronze we fall somewhere halfway between. This is the key to gating and risering. The risers and gates should be larger than red brass and tin bronzes, but not as large as would be required for manganese bronze. So the gates and risers requirements will fall somewhere between. As you get into it, you will soon realize that the solidification range of any metal or alloy is the key to gating and pouring.

ALLOY FORM

Various grades of silicon brass and bronze can be purchased in ingot form and also as casting shot. Casting shot is the alloy in the form of beebees.

Because silicon is quite popular with craftspeople and jewelrymakers for various small, centrifugal lost-wax castings, where the castings are in the weight range of only a few ounces to minus one ounce, the melting is done in these small lots with a torch in the crucible mounted on the arm of the casting machine.

You would have one big problem trying to reduce a 20-pound-plus ingot to small portions. With the shot, it is a simple matter to simply weigh out what is needed. The price of the alloys by the pound in shot form is quite greater than the per-pound price of ingots, but less expensive than trying to chew off a hunk with your teeth or a hacksaw.

Due to the simplicity of silicon bronze alloys and the availability of a wide choice of silicon/copper hardeners, it is a simple alloy to make up as needed by the foundry. Of course, if you pour a great deal of these silicon alloys, ingots might be the way to go. Scrap silicon bronze is not too hard to identify by fracture. The chocolate color and fine grain structure is usually a dead giveaway. The color of the polished surface looks very much like 14 karat gold. Asarco

154

ingots go by name Herculoy Regular, Herculoy Gear, Everdur 1000 and Tombasil (silicon brass). All have a common ingot number.

MOLDING SAND

Castings up to 100 pounds in weight can be made in the same sand as tin bronzes or red brass. When casting in green sand with Si bronze, it is important to keep the moisture below 6 percent. You will find that castings poured into green sand molds that are tempered with 6 percent or more moisture, when a surface which was in contact with a mold surface of 6 percent plus moisture when machining, you run into what appears to be a nonmetallic red brick dust material. Unless you have enough meat to machine this away to clean metal, you have lost the casting.

This condition will not be found in castings cast in green sand with a percent moisture below 6 percent. Castings over 100 pounds should be cast in dry-sand molds (no-bake, etc.). You can go up to several hundred pounds in a skin-dried mold if the skin is dried back far enough say 3/8 of an inch to 1/2 inch in depth and the mold is poured shortly after skin drying. If the mold is allowed to set too long prior to pouring, the moisture in the sand behind the skin-dried surface will migrate back to the surface defeating the purpose of skin drying. The best bet is go with a dry-sand mold.

FACING

Facing the mold is not usually done when casting silicon bronzes. The metal presents a good, smooth surface over a wide range of molding sand grain sizes.

GATES AND RISERS

With the recommended proportioning of risers of the diameter/height, the factor of 1 2/3 is commonly practiced. If you figure you need a 6-inch diameter riser, it should be 1 2/3 × 6 or 10 inches in height. An 8-inch riser would be 1 2/3 × 8 or 1.66 × 8 = 13.28″ in height.

Let's look at what a correctly designed riser should look like. From our example of a 6-inch diameter riser we used a height factor of 1 2/3 and came up with 10 inches for the riser height. See Fig. 8-1.

As shown in Fig. 8-1, you can see this riser would be a tough baby to remove. So let's neck it down with a neck core. A neck

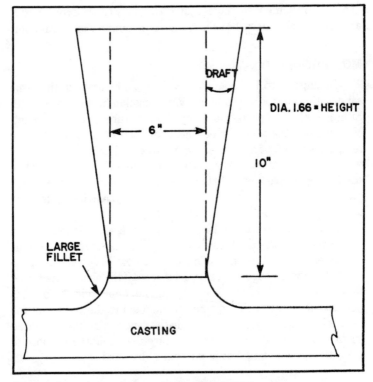

Fig. 8-1. A good riser design ratio for silver bronze.

core, also called an insulating core, is a dry-sand core that necks down the riser at its contact point, reducing the size of the contact point making riser removal much easier and simple. It also gives you a clean-cutting location, preventing you from cutting into the casting when removing, and leaves only a nub to be ground off. The core gets red hot on both sides. Not being too thick, it does not chill the riser neck to any degree. Thus the neck remains liquid, allowing the liquid reservoir above the neck to do its job of feeding the casting. See Fig. 8-2.

The same riser shown in Fig. 8-1 can be attached to the casting by a neck core. See Fig. 8-3.

Of course, you can insulate the riser with a sleeve or add an exothermic compound to the top and further reduce the riser diameter. Leave the height alone. Hydrostatic pressure is very important to the feeding of castings.

I have seen many castings lost due to too shallow of a cope depth above the casting. Some castings requiring risers for feed

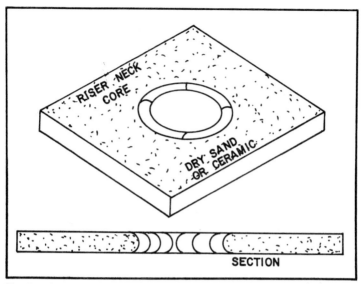

Fig. 8-2. A riser neck core is used to facilitate riser removal.

could be poured without risers if enough sprue height is available. See Fig. 8-4.

Let's look at a classic example of pouring a huge heavy casting that would require a very large riser. If you go up in height far enough, you can build up enormous hydrostatic pressure and ac-

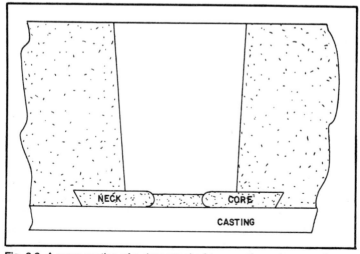

Fig. 8-3. A cross section of a riser attached to a casting using a neck core.

157

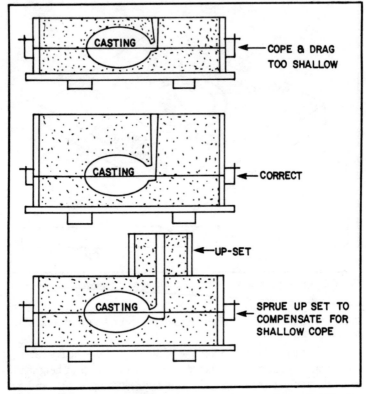

Fig. 8-4. Don't fudge on sprue height; you need the pressure to hold up the casting.

complish the same thing as if you had huge risers. The pressure does the trick the same way as if you were to centrifugally cast the item.

Figure 8-5 shows the gating used to cast a large gear blank via hydrostatic pressure with no risers. Mr. F.W. Paine, of San Francisco, specialized in large brass and bronze castings by this method. Sprue heights of up to 22 inches above the cope face of the casting were used.

Years ago we had a job to cast a very large nut (300 pounds plus). After two failures with risers, we tried Mr. Paine's method. With no risers whatsoever, we made several that when machined were perfect with no evidence of shrinkage porosity whatsoever. The gating had (as shown in Fig. 8-5) a 1-inch-diameter sprue that extended 24 inches above the top of the casting to a horn gate that was 1/2 inch in diameter at the casting-contact point.

With heavy work, if the design permits, the casting can be extended and then cut off to the correct size (length). See Fig. 8-6.

If you have a casting that is flanged on one or both ends, you don't have sufficient contact point for a riser. For feeding, you have to pad the riser to the end of the flange. Otherwise the necking down of the riser by the flange will freeze before the riser has had time to complete its work. See Fig. 8-7.

With silicon-bronze gating, you must use large-radius fillets, a minimum of 1/2 inch and better yet 1 inch on all gating components. Casting design should also have large fillets at wall junctions and section changes. Designers will fight this rule tooth and nail. Their objection to these large fillets (where a sharp right angle is desired or after the casting is machined) is where the rub comes. This large fillet radius is provided to avoid stress concentration at these junctures. See Fig. 8-8.

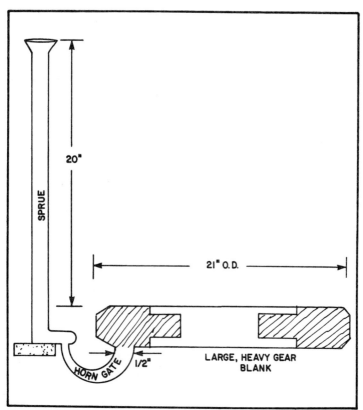

Fig. 8-5. Mr. Paine's high-pressure gating system for large, chunky jobs.

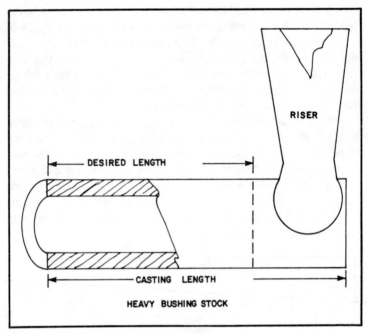

Fig. 8-6. A simple solution to riser removal.

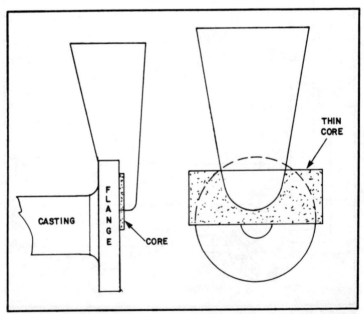

Fig. 8-7. The correct method of padding a riser to a flange, assuring proper feed.

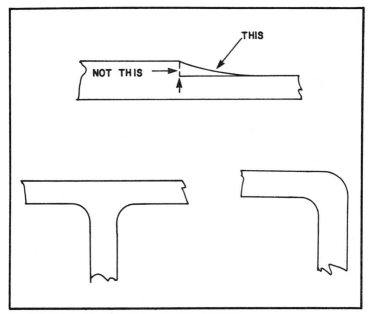

Fig. 8-8. Don't skimp on fillets with silicon bronze.

CORE PRACTICE

Good open sharp sand cores that are not too high in hot strength is all you need. The core must be able to collapse soon enough because these alloys are hot short at a red heat. Should the core be too strong and offer enough opposition to the metal's movement during the period where the casting is going through its red-heat period in cooling, you can run into hot tearing fractures and cracking. This is especially so if the core is almost completely surrounded by metal such as a hollow bushing stock or a manifold of some sort. Because a smooth cast surface is a characteristic of silicon bronze, usually no core wash is used.

MELTING

Silicon bronze and silicon brasses can be melted in crucible and open-flame furnace alike with ease. You can melt silicon bronze in the cupola and remelt over and over with no loss.

Melting silicon bronze/brass in an open-flame furnace or the cupola repeatedly does not seem to affect the physical properties. Simply melt in a slightly oxidizing atmosphere to 100 °F above the desired pouring temperature (pull the heat but don't soak). When it reaches the pouring temperature, pour with the ladle or crucible lip close to the sprue; keep the sprue choked.

FLUXING AND DEOXIDIZING

Normal melting with clean material requires no flux or deoxidizing treatment. With silicon copper alloys, you should avoid charcoal as covers and riser insulators. If the metal is melted improperly (reducing atmosphere) or allowed to absorb oxygen and/or hydrogen, it can be deoxidized with phos. copper as you would with a red brass. If the poured test cock fails to shrink, indicating a gassy heat, you can simply pig the melt and then remelt it under the proper conditions and melting practice.

If you add 1 percent nickel to silicon bronze, you can increase the tensile strength by as much as 40 percent and the yield strength by as much as 175 percent (quite a nice increase). If the tensile strength is 50,000 PSI as cast with 0 percent Ni, then with 1 percent Ni you are looking at 70,000 PSI (a gain of 20,000 PSI).

With a yield strength (as with 1 percent Ni), your gain is about 31,500 PSI if the yield was approximately 18,000. With the addition of the nickel, you must heat treat the castings by heating the castings to 1560 °F, quench them in room temperature water, and draw back by soaking the castings at 810 °F for one hour for each 1 inch of casting thickness. Of course, the nickel increases the hardness and reduces the percent of elongation in 2 inches.

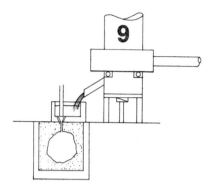

Phosphor Bronze

Phosphor bronze is simply defined as any copper tin alloy, with or without lead, that has a residual phosphorus content of not less than 0.10 percent.

The phosphorus exists in the alloy mainly as Cu_3P that freezes at 1300 °F. It is a very useful alloy but quite tricky in that it acts like platinum (when solidifying it is either liquid or solid). It does not pass through a solidifying range of crystallization (a thickening stage) like most alloys. You can say the phosphor bronze has a zero solidification range. If your pyrometer is off, you could find yourself in a real fix. You can be pouring a casting and have the metal simply go from a liquid to a solid in the crucible or ladle, and then you are out of business.

This fact, coupled with the extreme fluidity and lack of an oxide skin, causes the metal to eat into the mold and core sand. This penetrates into the very pores of the mold and core surfaces unless a provision is made (core and mold washes) to prevent this. You will simply get a casting covered with a fuzzy, very rough surface of a mix of metal and sand combination. You would suspect that, with the residual phosphorus in the melt, you would have little or no problem with gassing. This is not so. Phosphor bronze is very sensitive to gassing, and it must be melted properly under a slightly oxidizing atmosphere.

Phosphor Bronze Ingot

I would not order a phosphor bronze ingot first off. It is a special ingot that will cost you an arm and a leg. Your best bet is to add the phosphorus to your own specifications by using phos. copper shot. You can melt and add the required phosphorus and pig or make up the melt and pour the castings.

Phosphor bronze is not usually a special alloy; all grades of bronze can be converted into it. Its original name was steel bronze, and it was a straight tin bronze: 92 percent Cu, 8 percent Sn, and simply deoxidized with enough phos. copper to leave a residual percent of phosphor of 0.25 percent.

Standard phosphor bronze for bearings is usually 80 percent Cu, 10 percent Sn, 10 percent Pb with 0.25 phosphorus, or simply our old high-leaded tin bronze ingot #305, 80-10-10 with a shot of phosphorus.

Wrought phosphor bronze for making springs is 95 percent Cu, 5 percent Sn, and 0.25 Phos. Another spring phosphor bronze is 90 percent Cu, 10 percent Sn, and 0.25 percent phos. Phosphorus gives the metal a high resistance to fatigue failure.

ALLOYING EXAMPLE

Table 9-1 shows the basic allow of 89.8 percent Cu, 10 percent Sn, and 0.20 percent Pb, as cast, and cast with the addition of 0.5 percent phosphorus. By adding only half of a percent of phosphorus, you have gained in elongation, tensile strength, yield strength, and you have lowered the pouring temperature.

PHOSPHIDE MIGRATION

The eutectic is that portion of an alloy that remains liquid longer than the other component parts of the mixture. The upper part of the casting feeds the lower part with the eutectic, which seeps downward between the dentrites (solidified crystals). The riser, in turn, feeds or replaces the eutectic that has seeped down from the

Table 9-1. Alloys.

Alloy:	89.8-10-0.20 As cast 0% phosphorus	Alloy:	89.8-10-0.20-0.5% Phos.
Casting temp.:	2000°F	Casting temp:	1850°F
Color of fracture:	Light Buff	Color of fracture:	Gray
Tensile strength:	44,875 PSI	Tensile strength:	52,125 PSI
Elongation % in 2″:	41%		
		Elongation % in 2″:	42%:
Yield strength:	23,000 PSI	Yield strength:	24,375 PSI

top of the casting. That way you confine any porosity to the risers (natural shrinkage porosity). This is easily accomplished with an alloy that has a reasonable solidification range, such as red brass, that has a spread of up to 300 °F (1840 °F to 1540 °F).

With the phosphor bronze with its 0 °F solidification range, you have a problem. In obedience to a natural law as the eutectic drains downward into the lower voids, you leave behind voids between the faster freezing components that were formerly occupied by the sinking eutectic.

In an effort to make the risers feed these upper voids, the foundryman dealing with a "0" or very close range of solidifying often resorts to pouring extremely hot. In this way, this loss of eutectic metal in the top of the casting is compensated by the absorption of fresh, hot metal from the gates and risers (feed metal). Because phosphorized metal (phos. bronze) is so fluid, it will eat into the sand. At these higher pouring temperatures (especially with a high-lead, high-tin phosphor bronze), a phosphide migration will take place.

The phosphide migrates to the casting skin, where it displaces the lead, leaving you with a casting with a hard phosphide skin. The skin is extremely hard and abrasive, making the casting next to impossible to machine.

See Fig. 9-1. This phosphide (the suffix-*ide*) denoting a binary compound is usually a compound of phosphorus and copper.

Here is where you can get into trouble with phos. copper as a deoxidizer of red metals. When deoxidizing red brass, semi-red,

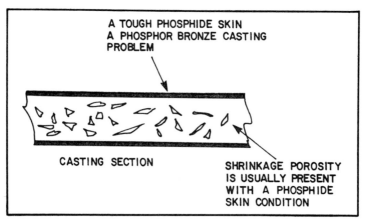

Fig. 9-1. Phosphide migration is an often misunderstood defect when casting high-phosphorus bronzes.

Table 9-2. Phosphorized Bronzes.

For Sand casting:				
% Phosphorus	.40	0.20	0.20	0.20
% Tin	18.25	7.20	10.00	10.00
% Lead	0	10.00	10.00	0.25
% Nickel	.50		.50	0.25
% Zinc	.25		.75	0.25
% Iron	.10		.15	
% Antimony	.15		.15	
% Copper	80.35	82.6	77.9	89.1

high-lead, high-tin alloys, etc., and you use to much, then you wind up with a phosphorus bronze or a casting with a copper/phosphorus phosphide skin. This problem crops up now and then with red and semi-red brass castings, puzzling the founder or machinist, only to be tracked down to the melter who is dumping the phos. copper in the melt by the fistful instead of weighing it up by the ounce. One ounce of 15 percent phos. copper per 100 pounds of metal is equal to 0.01 percent of phosphorus in the metal. This figure can be used to calculate for your desired residual phosphorus content when making up a phos. copper alloy.

If you examine your phos. copper shot, crush a small bead and you will see that it is a gray, very brittle material. This is phosphide. If you come up with a casting with a grayish, tough, abrasive skin, this is your copper phosphide.

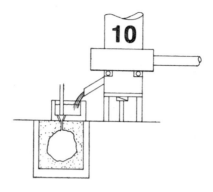

Aluminum Bronze

Aluminum bronzes form a family of tough, high-strength bronzes with good wearing qualities, hardness, corrosion resistance, and good fatigue resistance. Aluminum bronze castings are well suited for service at temperatures up to and including 750 °F. Some aluminum bronze alloys will respond to heat treatment.

They are tough alloys to cast due to drossing (foaming) if the gates are not designed in such a manner as to prevent turbulence of the metal as it enters the mold and goes in the mold. The gating should be quite simple displacement. Gate as you would with manganese bronze.

With aluminum bronzes, you have a very narrow solidification range from as low as a 6 degrees spread to the high of 30 degrees. Compare this with the other alloys I have covered and you will find quite a big difference in range (other than phosphor bronze with "O").

I don't mean to harp on the solidification ranges of various alloys but, as I stated, this is the key to gating a particular alloy. From a high of a 30-degree range to a low of a 6-degree range for aluminum bronzes, you know at once that these alloys are going to have extreme piping and are going to require very large risers to simply feed metal. When you have only a 6-degree range of temperature drop in which to get the casting supplied with feed metal, and the risers are not large enough to remain liquid long

167

enough, you have a problem. The volumetric shrinkage of the aluminum bronze alloys is only 3/16 of an inch per foot (the same as red brass). The problem is the short solidification range from liquid to solid.

Let's look at it this way. We have a semi-red brass casting alloy with a solidification range of 300 °F. We have provided a riser to feed the casting. We pour the casting and in the mold it starts to solidify at 1840 °F and is dropping in temperature at the rate of 1 ° per second, and it will be solid at 1540 °F. We have a 300-degree spread with which to work. At 1 ° per second, we have 300 seconds or five minutes of feed time. No problem. If we have a 6 ° spread, we have only six seconds in which the casting can be fed. Of course you can use chills, insulating sleeves, and topping compounds to help in the reduction of the number and size of the required risers. Large, carefully placed risers are usually required for all aluminum bronze castings other than small, even-walled castings.

As with any alloy or metal with a narrow solidification range, casting design is very important. You must have large fillets. Avoid heavy and light section abrupt changes, isolated heavy sections or difficult or impossible-to-feed situations.

Another problem with aluminum bronze castings is that you must shake out the castings as soon as they solidify (when they are solid enough to handle without distortion). If the castings are allowed to cool slowly in the mold, they will self-anneal. This "self-annealing" phase change causes embrittlement in the casting. If this is allowed to happen, it can be rectified by heat treating the casting. Aluminum bronze castings with a high nickel content do not self-anneal if cooled slowly.

I would put a shake-out time for aluminum bronze castings at no later than 30 minutes after pouring. This gives you a bench mark to go by. When you shake out at or before 30 minutes, this also means you must scrape the sand off of the casting immediately to facilitate air cooling.

Let's look at this self-annealing embrittlement. To sight an actual case, here are two test castings poured from the same heat. The No. 1 casting was allowed to cool in the mold whereas the No. 2 casting was shaken out as soon as possible, and the sand was scraped from the casting. The No. 1 casting flew apart and shattered as soon as pressure was applied by the breaking press. The No. 2, which was shaken out at a bright red heat, had an elongation percent in 2 inches of 35 percent and a tensile strength of 80,000 PSI.

HEAT TREATMENT

Some aluminum bronzes are heat treatable and some are not. For those that are heat treatable the process is quite simple. The castings are shaken out at a red heat. When cool enough to handle, they are placed in the heat-treat furnace and heated to the solution temperature for one hour per inch of casting thickness. The casting is then rapidly quenched in room-temperature water. The casting is then returned to the heat-treatment furnace and heated to the annealing temperature for one hour per inch of thickness, removed, and again quenched.

The time in the furnace at the desired temperature is known as the soak temperature. Generally, a typical heat treatment for a 3-inch-thick aluminum bronze casting would be like this: Heat the casting from 1600°- to 1650°F, soak for three hours at this temperature and quench in water. Reheat from 1100 to 1150°F and soak at this temperature for three hours (1 hour per inch in thickness) and quench in water. You have the two stages of heat treating, the solution soak and quench, and the annealing temperature and quench. These are general guidelines. Under the various aluminum bronze alloys, I will give the exact solution and annealing temperatures recommended for each alloy.

LET'S TALK ABOUT IT

It is not enough to simply give the physical side of heat treating; you must understand why it is done as well as what happens. Let's go back now and redefine a few things.

Eutectic Mixture. The eutectic mixture is solid solution of two or more substances having the lowest freezing point of all the possible mixtures of the components.

Eutectic Point. The eutectic point is the minimum freezing point attainable, corresponding to the eutectic mixture.

Coring. Coring is the segregation of low and high melting compositions. Often called dentritic segregation, coring is usually caused by very slow cooling upsetting the equilibrium conditions.

Isomorphous. My reference here in regards to isomorphous is to an alloy with complete solubility of the components in both the liquid and solid state.

Isomorphism. Isomorphism is the state in which two or more compounds from crystals are of similar shape.

Let's look at it this way. Some aluminum bronzes are isomorphous alloys, and therefore they are subject to coring when cooled slowly. This segregation of the eutectic solution would then cause a discontinuation of the structure as a whole, giving you a weak fragile mass.

Say you were going to glue or stick a bunch of ball bearings together with wax, and you did this in such a fashion that the wax was segregated throughout the mass. You would have a nonhomogeneous mix, loose-knit structure as a whole. See Fig. 10-1.

If you were to heat this arrangement until the wax melted (the wax represents our eutectic), the wax would then migrate throughout the mass of balls by capillary action until you would have a much more homogeneous mix (uniform composition throughout). Then you dumped the mass into cold water to freeze. The wax would represent the first stage of the heat treat. See Fig. 10-2.

Heat to the eutectic solution temperature and chill to arrest the now evenly dispersed eutectic. Now everything is under the stress of the sudden chilling. Therefore, you must stress relieve the whole mass by heating it up again, but this time keep the temperature below the eutectic solution temperature. This lets everything ad-

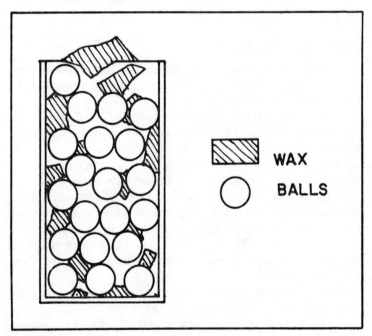

WAX

BALLS

Fig. 10-1. An analogy.

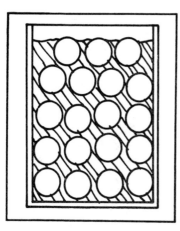

Fig. 10-2. An analogy.

just to a better state of equilibrium and chill again. This represents the second stage of the heat treat.

If, at our first stage of heat treat, you did not chill and allow the mass to cool slowly, you would have a mass segregation of the wax again. In many cases, what is known as autogenous heat treatment is all that is required to obtain the desired physical results. The casting is removed from the sand as soon as possible while it is at a bright red heat. The sand is scraped off and the casting is allowed to cool on the foundry floor, preferably in a draft (in a doorway). You can hasten the cooling by a water spray. If you spray with water, use good judgement to avoid casting warping.

I have compared ball bearings and wax to metal crystalline structure so let's look at the metal. Figure 10-3 shows three different micrographs representing three different conditions. As shown in A of Fig. 10-3, coring precipitation of large particles of the eutectic phase is caused by slow cooling. As shown in B of Fig. 10-3, the eutectic phase is dissolved. What you have is a supersaturated condition. This is accomplished by dissolving the random precipitation (shown in A of Fig. 10-3) by solution heat treating the first stage of heat treat. As shown in C of Fig. 10-3, three is a finely divided precipitate throughout the grains and the grain boundaries. This is accomplished by the second phase of the heat treat where you reheated the casting at a temperature below that of the solution temperature, soaked it at this temperature, and then quenched.

Not all heat-treatable alloys act the same. With some alloys, the precipitate is principally at the grain boundaries while others the precipitate is distributed uniformly throughout the grains. In

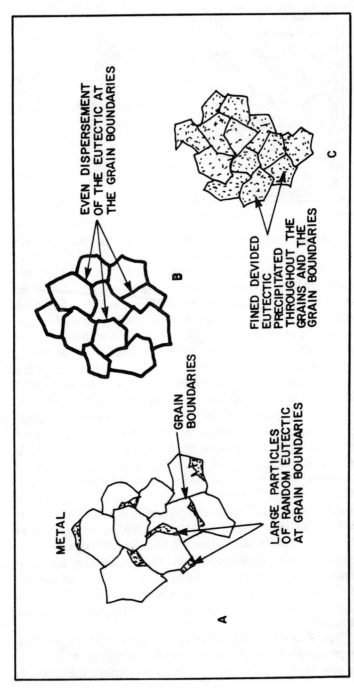

EVEN DISPERSEMENT
OF THE EUTECTIC AT
THE GRAIN BOUNDARIES

B

FINED DEVIDED
EUTECTIC
PRECIPITATED
THROUGHOUT THE
GRAINS AND THE
GRAIN BOUNDARIES

C

METAL

GRAIN
BOUNDARIES

LARGE PARTICLES
OF RANDOM EUTECTIC
AT GRAIN BOUNDARIES

A

Fig. 10-3. What happens to the eutectic during heat-treat stages.

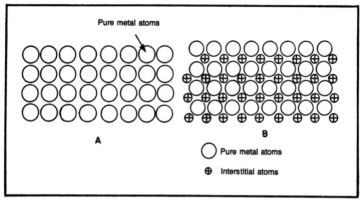

Fig. 10-4. Atom arrangement (pure metal versus an alloy).

some cases, it is selective on certain crystallographic planes within the grains.

SATURATED SOLUTION

With the saturated solution shown in A of Fig. 10-3, you have a condition called interstitial (defined as an additional atom or ion situated between the normal sites in a crystal lattice). Let's look at the crystal lattice of a pure metal (see Fig. 10-4).

As shown in A of Fig. 10-4, let us say the balls are pure metal atoms, and in B of Fig. 10-4 we have a solid solution of interstitial atoms. As shown in B of Fig. 10-4, we have an array of metal atoms represented as uniform spheres with smaller foreign eutectic atoms placed in the holes between the larger spheres. With any attempt to apply a shearing or deforming force to the metal, these smaller interstitial spheres will jam the movement of the larger spheres (such as an obstacle such as a wedge in front of a wheel).

Only a small addition of a few tenths of 1 percent is required to increase the hardness and strength. See Fig. 10-5. It is only

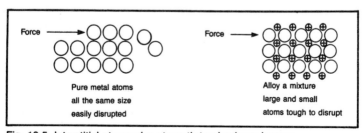

Fig. 10-5. Interstitial atoms give strength to aluminum bronze.

Table 10-1. Aluminum Bronzes.

Name & Composition	Properties	
Aluminum Bronze 9A 88% Cu 3% Fe 9% Al	Weight lbs/in^3	0.276
	Patternmaker's shrinkage	1/4" per ft.
	Solidification range	1913-1907°F
	Pouring temp. light work	2050-2250°F
	Pouring temp. heavy work	2000-2100°F
	Tensile strength	80,000 PSI
	Yield strength	27,000 PSI
	Elongation % in 2"	35%
	Brinell hardness (3000 kg)	125
	Machinability	50
	Heat treatment	No response
Aluminum Bronze 9B 89% Cu 1% Fe 10% Al	Weight lbs/in^3	0.272
	Patternmaker's shrinkage	3/16" per ft.
	Solidification range	1913-1904°F
	Pouring temp. light work	2050-2250°F
	Pouring temp. heavy work	2000-2100°F
	Tensile strength	**As cast** 75,000 PSI Heat treated 85,000 PSI
	Yield strength	**As cast** 27,000 PSI Heat treated 42,000 PSI
	Elongation % in 2"	**As cast** 25% Heat treated 15%
	Brinell hardness (3000 kg)	**As cast** 140 Heat treated 174
	Machinability	50
Aluminum Bronze 9C 85% Cu 4% Fe 11% Al	Weight in lbs/in^3	0.269
	Patternmaker's shrinkage	3/16" per ft.
	Solidification range	1900-1880°F
	Pouring temp. light work	2050-2250°F
	Pouring temp. heavy work	2000-2150°F

	Tensile strength	As cast 85,000 PSI / Heat treated 105,000 PSI
	Yield strength	As cast 35,000 PSI / Heat treated 54,000 PSI
	Elongation % in 2"	As cast 18% Heat treated 8%
	Brinell hardness (3000 kg)	As cast 170 Heat treated 195
	Machinability	60

Aluminum Bronze 9D
81% Cu
4% Ni
4% Fe
11% Al

Weight in lbs/in^3	0.272	
Solidification range	1930-1900°F	
Patternmaker's shrinkage	3/16" per ft.	
Pouring temp. light work	2050-2300°F	
Pouring temp. heavy work	2000-2150°F	
Tensile strength	As cast 100,000 PSI / Heat treated 120,000 PSI	
Yield strength	As cast 44,000 PSI / Heat treated 68,000 PSI	
Elongation % in 2"	As cast 12% Heat treated 10%	
Brinell hardness (3000 kg)	As cast 195 Heat treated 230	
Machinability	50	

Alpha Nickel Alum. Bronze
81% Cu
5% Ni
4% Fe
9% Al
1% Mn
Note here the manganese

Weight in lbs/in^3	0.276
Patternmaker's shrinkage	3/16" per ft.
Solidification range	1940-1910°F
Pouring temp. light work	2050-2300°F
Pouring temp. heavy work	2000-2150°F
Tensile strength	95,000 PSI
Yield strength	38,000 PSI
Elongation % in 2"	25%
Brinell hardness (3000 kg)	159
Machinability	50
Heat treatment	No response

necessary in some alloys that you have second, smaller atoms scattered around in the voids between the larger atoms to give you a very marked difference in physical properties. This is called interstitial solid solution. In some close-packed metal structures, not all of the interstitial holes are the same size. Some are not large enough to accept a smaller foreign atom.

What sometimes looks like a minor impurity in an alloy formula because of its percent is a decimal percentage. Don't let it fool you because it might be the most important element in the recipe.

Remember the marked difference in the physical properties between an alloy without any phosphorus and the same alloy with the addition of only a minor decimal amount of phosphorus? Phosphorus is a non-metallic element. The interstitial atoms do not have to be always metallic atoms or metallic compounds. Silicon (Si) is a widely used none-metallic material for an alloying agent in steels, aluminum, bronze, copper, and iron.

Pure aluminum has a tensile strength of 12,000 PSI, and it is soft and malleable. If you add 5 percent silicon, the tensile strength jumps by 5,000 PSI to 17,000 PSI. It is no longer soft and malleable.

Table 10-1 shows the basic five alloys used today. Three of them will respond to heat treatment. See Table 10-2.

The solution temp and the annealing temp in all cases is held (soaking time) for at least one hour per inch of casting thickness taken at the thickest spot. The quench water must be circulated rapidly or the casting must be agitated in the quench.

Unless you have ample circulation of the quench (cold water) or motion of the casting you could come up with different physical properties. Aluminum bronze has a "mass affect." The heavy sections will try to cool slower than the light sections. The trick is to try to cool the whole casting down quickly and as evenly as possible. It will pay you to get a good reference on the principles of heat treating or talk with someone in the game. It is quite common, if a foundry is producing a large tonnage of castings requiring heat

Table 10-2. Heat Treating Aluminum Bronze

Composition:	89% Cu, 1% Fe, 10% Al—Solution temp. 1625°F Annealing temp. 1200°
Composition:	85% Cu, 4% Fe, 11% Al—Solution temp. 1650°F Annealing temp. 1200°F
Composition:	81% Cu, 4% Ni, 4% Fe, 11% Al—Solution temp. 1610°F Annealing temp. 1200°F

176

Table 10-3. Alloy Ingots.

#415-9A-1	Min. %	Max. %
Copper	86.75	88.75
Iron	2.75	3.25
Aluminum	8.50	9.50

#415-9A-2	Min. %	Max. %
Copper	78.00	86.00
Iron	3.00	3.75
Aluminum	10.30	11.20
Manganese	0	3.50

treatment, to job this out to a heat-treating firm. For a now-and-again heat treat, you can improvise.

FOUNDRY PRACTICE

Aluminum bronzes require the same practices as manganese bronzes, with some minor modifications. Let's look at them.

ALLOY FORM

All the alloys in ingot form are designated as ingot #415 (followed by the grade). Ingot #415-9A would be 88 Cu, 3 Fe, 9 Al. Ingot #415-9B would be 89 Cu, 1 Fe, 10 Al, etc. Numbering systems have gotten so confusing and complicated that you need a metal numbering identification thesaurus a foot thick just on metals and metal alloys. Every smelter assigns a name to each of his products and his number. The Navy assigns its number and name, and the Army also.

Then you have the unified numbering system (UNS) and the copper alloy number (CA). The ASTM and SAE people have their numbers. And every once in a while it all changes or the society of bowlegged molders decide to come up with new names and numbers. Let's look at alloy ingot #415-9A-1. The plot thickens. See Table 10-3. You could go on forever.

The same thing applies to various heat treatments recommended for the aluminum bronzes that respond to heat treatment. You will find a wide range of recommended solution temperatures and annealing temperatures. The annealing temperature is very often referred to as the drawing temperature. You will find that the specific solution temperature and the annealing temperature for each alloy will have to be worked out to give you the combination of strength, hardness, and ductility best suited for the intended application and specification of the particular casting or castings.

Very generally, with any alloy with 10 percent or more aluminum content (the ones which respond to heat treatment), you are looking at a range of from 1550 °F to 1750 °F for the solution temperature and 800 °F to 1200 °F for your annealing or drawing temperature.

In very many cases, the autogenous (self-generated) heat treatment precludes heat treatment by other methods. So check before you jump.

Most foundries purchase ingot for melting. You can, however, purchase a hardener that is 70 percent pure aluminum and 30 percent Armco iron. Armco is the tradename for an ingot iron that is 99.94 percent pure (with a carbon content of only 0.013 percent and manganese of only 0.017 percent). It can be purchased in shot form for making up alloys. Plastiron, by Henry Disston & Sons Inc., is another grade of pure iron.

The procedure is quite simple; figure the percent of hardener required, melt down the copper, and add the required percent of preheated ALFe hardener to the copper in small lots. Stir each in well until the correct amount is dissolved in the copper.

Because the aluminum iron hardener is quite a bit lighter than copper, so it wants to float (thus the reason for stirring it in well). I have seen some very beautiful aluminum bronze castings produced in a backyard foundry from homebrew aluminum bronze. The hardener was made by dissolving clean, soft-iron baling wire in a heat of aluminum beer cans. The hardener was then poured off into cold water to granulate it.

The copper consisted of clean #1 heavy copper. It is not as tough as it looks, and you can beat the high price of ingot and make only what you need. To prevent contamination of your aluminum bronze heats, the turnings, small spills, skimmings, grindings of aluminum bronze previous heats should not be used in the melt. They should all be saved and sold as scrap to the smelter.

MOLDING SAND

Grain size is 100-160, clay percent is 10-20, permeability is 15-50, green strength is 5-12 PSI, and moisture is 4 to 6 percent. Just about any nonferrous sand, natural or synthetic, will do a good job. The aluminum skin that forms on the metal, if not broken by turbulance, will give you a smooth casting even in more open sand than is normally used in nonferrous work.

FACING

The aluminum in the alloy gives it a very high surface tension similar to silicon bronzes. Therefore, the use of a facing is usually not required. Aluminum bronzes will give a much smoother as cast surface than a red brass or tin bronze cast in the same sand.

GATES AND RISERS

Gate as you would with manganese bronze. Pour close to the mold, and fill as quickly as possible with a minimum or no turbulence by simple displacement (inverted horn). See Fig. 10-6. Risers, when required, need to be somewhat larger than those for manganese bronze.

Note: Usually if you have a number of castings to pour and they are not too large, you can pour a sample mold without risers, (just a simple horn gate and let the casting shrink, etc.). This sample will give you a very clear picture of where and how many risers and chills are required, plus the size necessary. This practice is always valid regardless of the alloy.

CORE PRACTICE

You can get away with just about any good basic oil sand or no-bake core mix. Unlike most red metal alloys, aluminum bronze does not have a crucial temperature where it could or would not tear. In fact, it is actually forgetable at any temperature. So core and mold excessive hot strength is not crucial. By this I don't mean to use rock-hard cores with low collapsability. What I mean is that, as the casting shrinks during solidification, if the core or mold restricts this movement the casting will not crack, rupture, or tear apart. See Fig. 10-7.

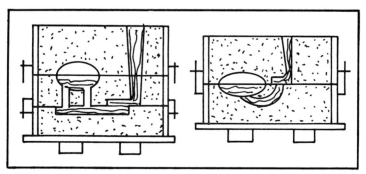

Fig. 10-6. Fill aluminum bronze molds by simple quiet displacement.

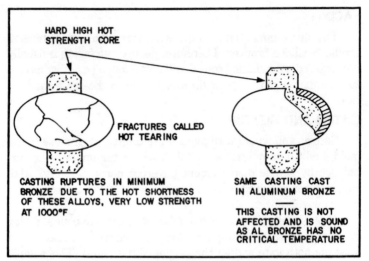

Fig. 10-7. Comparison of Al bronze with Mn bronze (hot shortness).

MELTING

With a crucible furnace, do not stir; simply heat to 100 degrees above the selected pouring temperature, skim, and pour.

FLUXING AND DEOXIDIZING

You usually do not need any type of fluxing or other treatment if you melt with the burner adjusted slightly oxidizing. You can cover the crucible with an old crucible bottom or a purchased crucible cover.

If a test cock indicates you have to do some deoxidizing, the best deoxidizer for aluminum bronze is pure aluminum. A small amount goes a long way.

Electrode or EDM carbon makes excellent chills for aluminum bronze. Shake castings out hot to prevent grain growth.

IMPURITIES

Lead is the biggest offender. Keep lead below .25 percent (preferably to "0" percent). Lead will discolor the surface. Keep tin below 0.5 percent. Silicon is detrimental in amounts above 0.05 percent because it will cause embrittlement. Zinc up to 0.75 percent seems to have no effect.

The biggest problem is getting aluminum oxide skins into the melt due to stirring and turbulent gating.

180

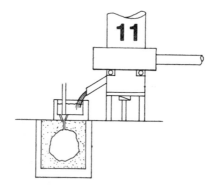

Other Red-Metal Alloys

I have covered the most common red-metal alloys, but the woods are full of other alloys that you could call specialty metals. I could go on and on until we would have a huge volume.

I would prefer to give enough information on the more common specialty alloys whereby the interested founder could take it from there. Various smelters and metallurgists worldwide are continuously coming up with new alloys and variations of old ones. Some that were commonplace 50 years ago have all but disappeared; they have been replaced by the newer ones. Ever hear of Alfendide metal 59 percent Cu, 30 percent Zn and 11 percent Ni or Allen Red Bronze 62.5 percent Cu, 7.5 percent Sn, 30 percent Pb? For every one alloy that leaves the scene, two new alloys take its place. In addition, a world of alloys that were common have been replaced with plastics.

CUPRO-NICKELS

Alloy UNS #C96,200, called 90-10 copper nickel, has a composition of 86 percent Cu, 10 percent Ni, 1.4 percent Fe, 1.0 percent Cb (Columbium, also called Niobium) a rare gray metal, and 1.0 percent Mn. Patternmaker's shrinkage is 3/16 of an inch per foot and pouring temperature is 2100°F to 2600°F.

Its sister alloy, UNS C96,400 is called 70-30 copper nickel. Copper nickel alloy has the composition of 67 percent Cu, 30 percent Ni, 1.0 percent Fe, 1.0 percent Cb and 1.0 percent manganese, a higher tensile strength alloy, patternmaker's shrinkage of 7/32 of an inch per foot, and a much higher pouring temperature (from 2550 °F for heavy work to as high as 2700 °F for light work). The nickel is what shoots the pouring temperature up so high. At 2700 °F you are looking at cast-iron temperatures. Therefore you are looking at cast-iron sand practice.

The basic uses for these alloys is for castings that are subjected to a salt-corrosive atmosphere near seawater or used for pipe fittings, valves, marine hardware, etc. They are very weldable and have excellent mechanical properties. They are deoxidized with 1/2 ounce of magnesium per 100 pounds of melt. You must tie the magnesium to a rod and plunge it to the bottom of the melt. Your sand should be low in moisture, high in green strength, and very refractory.

The solidification range for 90-10 is from 2100 °F to 2010 °F (or a 90-degree spread) and 70-30 has a range of from 2260 °F to 2140 °F (or a 120-degree spread). So you are looking at large risers and tough melting and pouring temperatures.

NICKEL SILVERS

Nickel-silver alloys are white in color and are often referred to as German silver. The silver is in reference only to the color. They have no actual silver content.

The leaded-nickel silver alloys are easily machined. They also have superior resistance to galling.

Deoxidizing these alloys requires a treatment of 5 ounces of 70-30 copper manganese shot, 1 ounce of magnesium, and 3 ounces of 15 percent phosphor shot. This treatment should be done approximately three minutes prior to pouring. The magnesium, of course, must be tied to the stirring rod and plunged to the bottom of the melt. Remember to use cotton string to tie the magnesium; 10 to 12 wraps will do it.

As shown in Table 11-1, there are basically four widely used nickel silvers. Because these alloys have a high solidification shrinkage, large gates and risers are normally required. The general foundry practice is similar to the cupro nickels (nickel bronzes).

The pouring range is from 2000 °F for very heavy work to as high as 2600 °F for very light work. The 25 percent, 12 percent, and 17 percent nickel alloys have a patternmaker's shrinkage of 3/16 of an inch per foot. The 20 percent Ni alloy has shrinkage of 1/8 of an inch per foot. These alloys do not respond to heat treat-

Table 11-1. Nickel Silvers.

12% Ni Silver	17% Ni Silver	20% Ni Silver	25% Ni Silver
56% Cu	59% Cu	64% Cu	66% Cu
2% Sn	3% Si	4% Sn	5% Sn
10% Pb	5% Pb	4% Pb	2% Pb
12% Ni	17% Ni	20% Ni	25% Ni
20% Zn	16% Zn	8% Zn	2% Zn

ment. Such alloys do have a wide range of uses as ornamental, decorative, plumbing, food-handling equipment, marine, and builder's hardware.

Nickel silver is often referred to as nickel brass. The ASTM defines nickel brass as any copper alloy with over 10 percent zinc and enough nickel to give the alloy a white color. Nickel bronze is defined as a copper alloy with over 10 percent nickel, with the zinc less than the nickel and with the lead under 0.5 percent.

The leaded nickel bronzes are similarly defined except they contain over 0.5 percent lead.

One intersting alloy used to replace nickel silver is an alloy, produced by American Smelting & Refining Co., called Bronwite. Because this alloy contains little or no nickel, costs and melting temperature are reduced. This also eliminates some of the problems encountered with nickel alloys. The claim by AS & R Co. is that Bronwite is equal to nickel bronze as to color, lustre, strength, ductility, hardness, and it is a great deal more versatile because of its lower melting temperature 1550 °F.

Let's look at the composition and properties of Bronwite. Composition is 58 percent Cu, 1 percent Pb, 20 percent Zn, 20 percent Mn, and 1 percent A1. Patternmaker's shrinkage is 5/16 of an inch per foot, solidification range is 1550 °F to 1505 °F, pouring temperature for light work is 1650 °F to 1850 °F, pouring temperature for heavy work is 1600 °F to 1700 °F, and tensile strength is 60,000 to 70,000 PSI. I am sure if you were to contact ASARCO they would give you the cost and details of this alloy. Other smelters produce similar alloys, plus a raft of other specialty alloys.

BERYLLIUM BRONZE

Beryllium (Be) is a steel-gray, lightweight, very hard metallic element formerly known as glucinum. It is the lightest structural metal known. Beryllium is not what you would consider a tonnage metal. Beryllium sheet and other forms such as tubing and powder rank almost as precious metals. Its chief commercial use is for hardening copper and nickel for beryllium bronzes. It is also used in fine, high-grade 18-karat gold jewelry as a hardener to enhance wearing qualities. The 18-karat gold, alloyed with a minor percent of beryllium, will have a Brinell hardness of up to 300.

Beryllium copper alloys are precipitation hardenable with tensile strengths up to 175,000 PSI and an elongation of 5 percent in 2 inches. The typical analysis is copper 97.4 percent, beryllium 2.25 percent, and nickel 0.35 percent. Beryllium alloys are produced by a number of companies under various trade names for various uses and operating conditions. Some common trade names of beryllium copper alloys in common use today are Berylco alloy 165, Beryldur, Berylco 717C (Viculoy, a group of alloys of Akron Bronze Co), Ampcoloy 91, Tuffalloy 55, Trodaloy, plus 15 or 20 more.

Beryllium is also alloyed with aluminum to produce special alloys. The Beryllium Corp. has an aluminum Beryllium alloy consisting of 62 percent beryllium and 38 percent aluminum, which has a 55,000 PSI tensile strength with a yield strength of 40,000 PSI and a service temperature of up to 800 °F.

Copper beryllium alloys have a wide commercial use in springs, bearings, valves, nonsparking tools (wrenches etc.), plastic molds, glass molds, turbine blades, and a large variety of strong mechanical parts. The relatively low pouring temperature and the excellent fluidity of copper beryllium alloys simplifies the foundry practice (giving castings good intricate detail). It is widely used by artists, jewelers, and sculptors. The preceding factors give you a greater freedom of design.

All beryllium copper alloys respond to heat treatment. Table 11-2 shows two of the most used beryllium copper casting alloys, which I will call alloy "A" and alloy "B." Let's look at the limits of "A" and "B" composition. The object is to hold the various constituents between given ranges. The following are properties of alloys "A" and "B."

- ☐ Alloy "A" as cast 75,000 to 90,000 PSI tensile.
- ☐ Alloy "B" as cast 95,000 to 110,000 PSI tensile.
- ☐ Alloy "A" heat treated 155,000 to 170,000 PSI tensile.
- ☐ Alloy "B" heat treated 170,000 to 185,000 PSI tensile.

From this you can see that alloy "B," when heat treated, is one very strong alloy. You realize a gain of 75,000 PSI tensile

Table 11-2. Copper Casting Alloys.

Alloy "A"	Alloy "B"
97.2% Copper	96.6% Copper
2.15% Beryllium	2.75% Beryllium
0.65% Cobalt	.65% Cobalt

Alloy "A"			Alloy "B"		
Beryllium	1.90% min	2.15% max	Beryllium	2.50% min	2.75% max
Cobalt	.35% min	0.65% max	Cobalt	.35% min	0.65% max
Silicon	0.20% min	0.35% max	Silicon	0.20% min	0.35% max
Iron not over	.25%		Iron not over	.25%	
Aluminum not over	.15%		Aluminum not over	.15%	
Tin not over	.10%		Tin not over	.10%	
Lead not over	.02%		Lead not over	.02%	
Zinc not over	.10%		Zinc not over	.10%	
Nickel not over	.20%		Nickel not over	.20%	
Chromium not over	.005%		Chromium not over	.005%	

Of course, copper makes up the remainder of the alloy.

strength by a simple precipitation hardening heat treat. The heat treatment used in the example of alloys "A" and "B" was as follows. The casting is heated to 1450°F for one hour per inch of the thickest section then quenched, followed by an annealing soak at 600°F.

PRECIPITATION HARDENING

I have discussed what happens when you heat treat aluminum bronze. Well, the same thing happens with copper-beryllium alloys. Very simply and basically, when you have a small proportion of a constituent in an alloy that can be put into solid solution at one temperature and then reprecipitated in a very, very fine state (very small particles) at another temperature, you have precipitation hardening. In many cases, with various alloys ferrous and nonferrous, this will take place by itself at room temperature over a period of time (self-hardening). Some alloys will gain a large amount of tensile strength in a few days.

With the correct heat treatment, you can accomplish hardening in a relatively short period. The precipitated substance is too fine to be seen under the microscope; its presence is shown by the improvement in the physical properties of the metal.

Copper can take up to 2.1 percent beryllium into solid solution. Tin bronze and leaded-tin bronze are not considered heat-treatable alloys. You will not gain any appreciable physical benefit by heat treating them because they will not precipitate harden.

The dentritic structure lends itself to discontinuity during formation; this fact is associated with weakness. This discontinuity is due in part to lack of feed, resulting in solidification shrinkage porosity.

If a tin bronze or leaded-tin bronze in this condition is annealed at 1000 °F a uniform grain can be established. I have seen valve castings that have leaked under test pressure, annealed at 1000 °F, and then retested as sound as a rock.

FOUNDRY PRACTICE

The molding practice is similar to any regular bronze. The advantage is that the pouring temperatures are low (1900 °F to 2100 °F), depending on the casting section and design. It is usually recommended that if the beryllium content is 1.90 percent to 2.15 percent you should not exceed a pouring or melting temperature of 2150 °F and you should not exceed 2000 °F if the beryllium is 2.50 percent to 2.75 percent.

With remelting scrap gates and risers, you will have some beryllium loss due to oxidation; an average is 5 percent and it can be in excess of 7.5 percent. Your ratio to virgin pig should be kept at about 30 percent scrap to 70 percent new ingot. Fine, natural-bonded sands are fine grain size 100 to 150, clay 10 percent to 20 percent, and permeability 17 to 40, moisture 4 percent to 6 percent (don't exceed 6 percent).

French sand or the equivalent will produce extremely smooth castings with sharp detail. The gate and risers would be like that recommended for tin bronzes because the shrinkage is less than manganese, aluminum, or silicon bronzes. As with any casting requiring a riser or risers, the gating should be designed whereby the riser or risers receive the last metal. Thus the hottest metal will be in the riser or risers. With small plated patterns, the runner should be in the cope with the gates backing into the cavities. See Fig. 11-1.

Use tin-bronze melting practice slightly oxidizing furnace atmophere. No flux, cover, or deoxidizer is necessary. You can melt beryllium bronze in an induction, rocking arc, rotary, or crucible furnace.

CORES

Core mixes suitable for tin bronzes and leaded-tin bronzes will

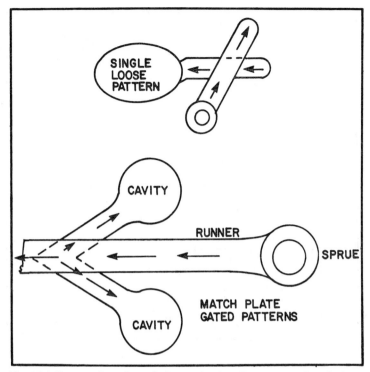

Fig. 11-1. Reverse flow of metal when gating beryllium bronze alloys for best results.

give good results with beryllium bronze. Use a good, open-core sand bonded with linseed oil or furan no-bake cores. A wash is usually not required. In some special cases, you might elect to wash the cores.

CHILLS

Where necessary, cast-iron chills that are free from rust and are dry work well.

MACHINABILITY

Beryllium bronze is machined like lead-free tin bronze, silicon bronze, and aluminum bronze. All are not what you would call a free-machining material. You have to work out the best tools clearance angles, speeds, and feeds. In most cases, it will take a liquid coolant (cutting fluid). Consult your local machine-tool dealer and supplier or call on a machinist.

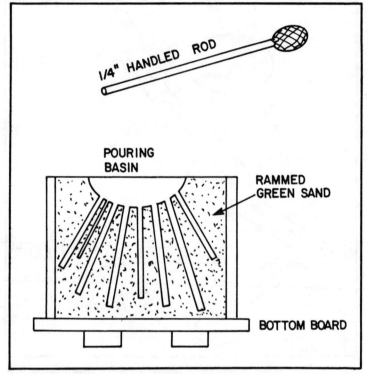

Fig. 11-2. A simple mold to cast chasing stock.

WELDABILITY

Beryllium bronze alloys are easily welded with beryllium/copper rods. These rods can be purchased with properties that approach the alloy being welded.

Often when welding a beryllium bronze casting to another or dissimilar metal component—such as you might do in producing a casting weldment—silicon-bronze rods or aluminum-bronze rods are used. Often in place of making a complicated casting, it is broken down into two or more simpler components and joined by welding and bolting.

What we used to do when casting beryllium bronze, silicon bronze, and aluminum bronze was to ram up a drag mold with a pouring basin, and, with a 1/4-inch rod, perforate the mold and cast welding rods from the same melt as the casting. In this way, we knew the rods matched the casting specifications. Stock made this way is called chasing stock. See Fig. 11-2.

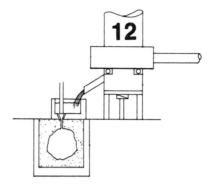

Commonsense Gates and Risers

You will find formulas, graphs, charts, etc., by some real wise fellows, that are supposed to eliminate the mystery of gating anything from a grate bar to a locomotive cylinder. A great many of the fellows who put out this hallooed information have never set foot inside a genuine foundry much less made a mold, cut a gate, or riser or (heaven forbid) ever actually gone the whole route and actually poured the mold they produced.

I'm not putting technology down completely. Each casting is different and represents its own problem and requirements. This precludes making any tried-and-fast formulation that would work over even a limited set of designs. There are many things to consider when gating a casting: casting weight, section thickness, design of casting, chosen alloy, fluidity of alloy, pouring temperature, alloys reaction (drossing), solidification range of alloy, and rigging.

It's a commonsense proposition. You have to consider all the factors. What you are trying to do is fill the mold with the metal as fast as possible, with the minimum amount of turbulence, and accomplish directional solidification whereby the gates and/or risers freeze last. all this you have to do without washing dirt, sand, or slag into the cavity. In some castings, the design is so bad that it is next to impossible to accomplish this.

In gating, a point that is overlooked many times is the removal of the gates and risers. In an effort to secure a sound casting, I have seen gating systems that cost more in labor and time to remove than the casting is worth. Some 88 percent of most casting defects can be attributed to improper gating because the gates were neither designed nor cut properly, and there was too promiscuous and unintelligent use of risers and sink heads. I could fill several volumes on gating and still not scratch the surface. Here are some general rules.

☐ The flow of metal should not meet with any obstruction on entering the mold.

☐ Usually gate the mold at the heavy section.

☐ Use the smallest sprue that will run the casting satisfactorily.

☐ For large, thin castings consider a simple drop gate.

☐ Consider bringing the metal in or near the bottom of the cavity.

☐ Horn gates can be very advantageous.

☐ Consider gating through an interior core.

☐ Round sprues are best.

☐ When you have a gate that proves satisfactory, do not experiment with it. Leave well enough alone.

☐ Touching up risers has saved many castings. It is a good practice and a practical safety measure.

Note: Touching up a riser is practiced when metal shows in the riser, the ladle is moved from the sprue, and hot metal is poured into the riser. This is done by pouring additional metal, in small quantities, into the riser until the riser or risers are filled to the desired height.

In some cases, the mold is poured through the sprue until the riser or risers are full. As they start to pipe (shrink down), you pour them full again from the top directly into the risers. Repeat as necessary.

With large manganese bronze and aluminum bronze, it is quite common—after the mold is poured—to bring up a hot crucible or ladle of additional metal. This is used to add fresh hot metal directly to the risers in order to compensate for shrinkage (supplying hot fresh metal for feeding).

Figures 12-1 through 12-35 show a few of some of the various methods of gating and risering castings.

190

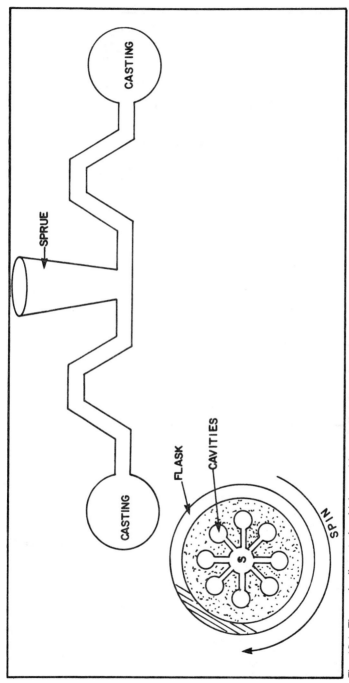

Fig. 12-1. The umbrella gate is widely used in centrifugal casting of green and dry-sand molds in round flasks.

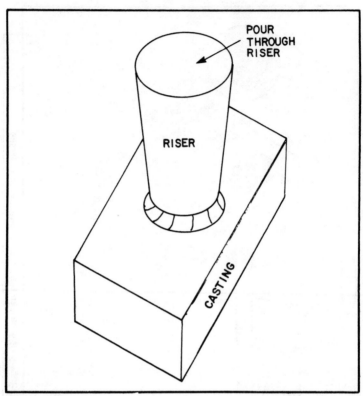

Fig. 12-2. Riser gate.

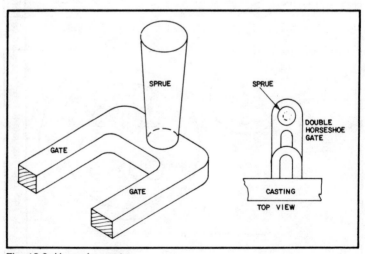

Fig. 12-3. Horseshoe gate.

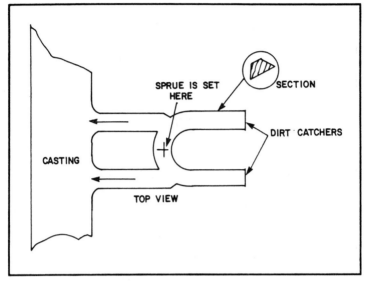

Fig. 12-4. Reverse horseshoe gate.

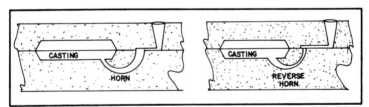

Fig. 12-5. Horn gate and reverse horn.

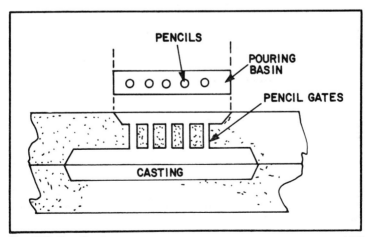

Fig. 12-6. Pencil gates.

193

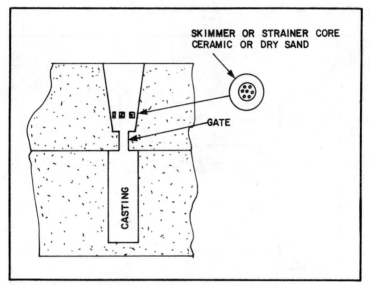

Fig. 12-7. Top gate and skimmer core.

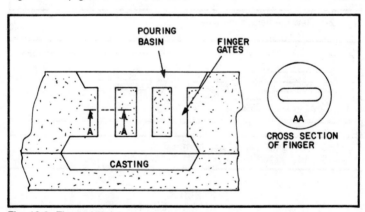

Fig. 12-8. Finger gate.

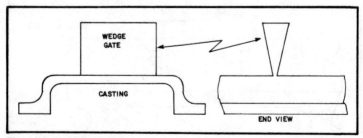

Fig. 12-9. Wedge gate.

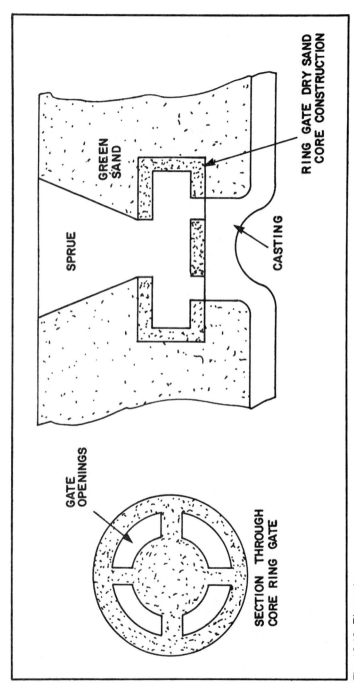

GREEN SAND

SPRUE

RING GATE DRY SAND
CORE CONSTRUCTION

CASTING

GATE
OPENINGS

SECTION THROUGH
CORE RING GATE

Fig. 12-10. Ring gate.

195

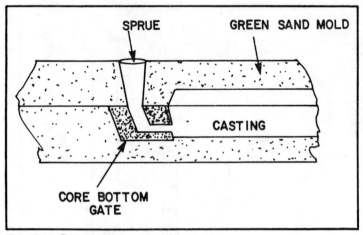

Fig. 12-11. Bottom gate.

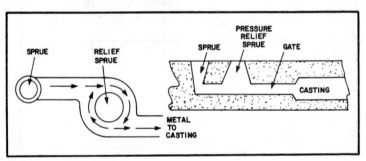

Fig. 12-12. Whirl gate.

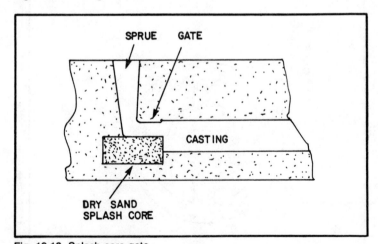

Fig. 12-13. Splash core gate.

196

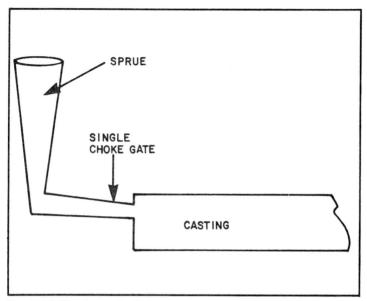

Fig. 12-14. Simple, single-choke gate.

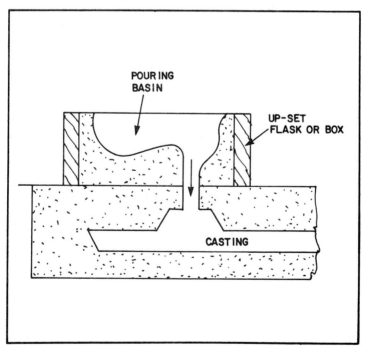

Fig. 12-15. Pouring basin top gate.

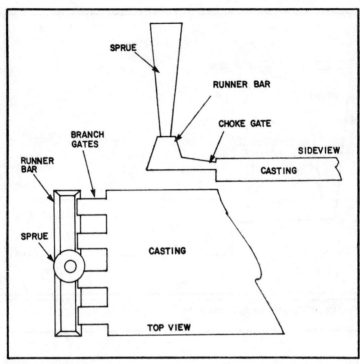

Fig. 12-16. Branch gate.

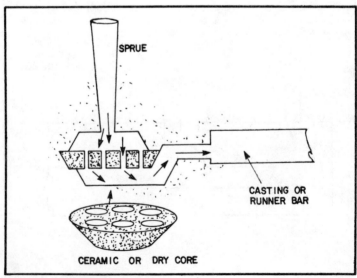

Fig. 12-17. Strainer-core gate.

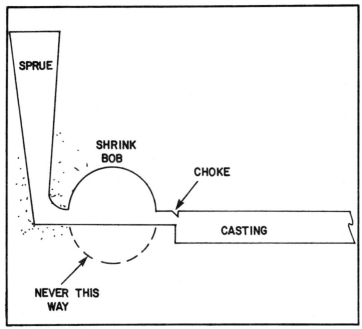

Fig. 12-18. Shrink-bob gate.

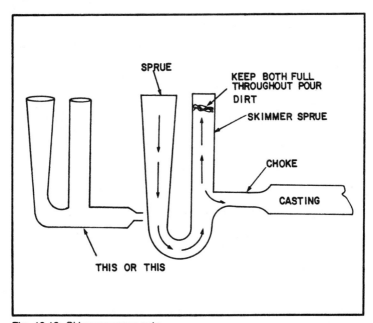

Fig. 12-19. Skimmer-sprue gate.

199

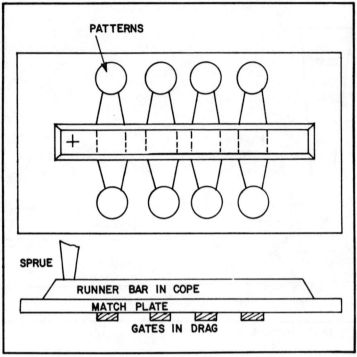

Fig. 12-20. Common multipattern match-plate gate.

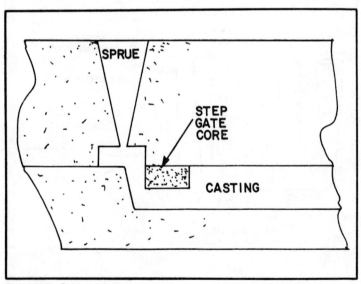

Fig. 12-21. Step gate.

200

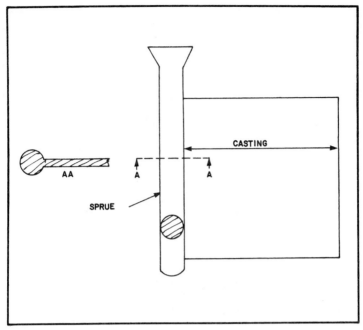

Fig. 12-22. French gate.

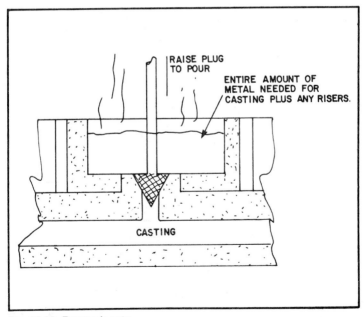

Fig. 12-23. Reservoir gate.

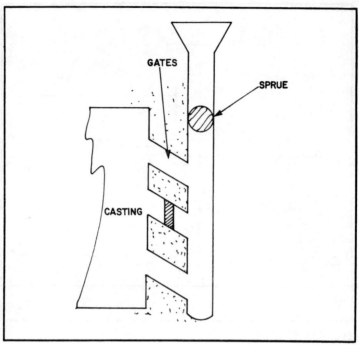

Fig. 12-24. Saxophone gate.

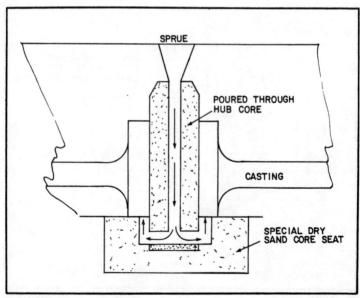

Fig. 12-25. Through core gate.

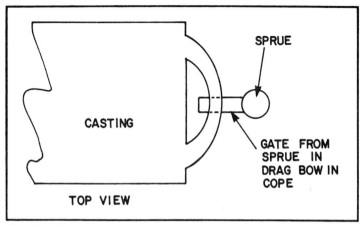

Fig. 12-26. Bow gate.

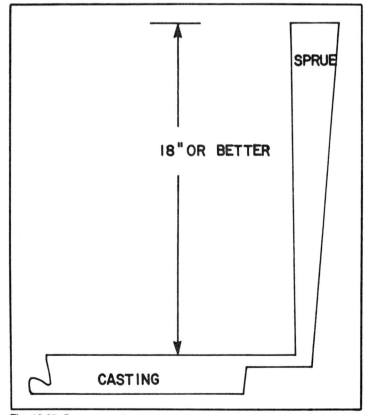

Fig. 12-27. Pressure gating.

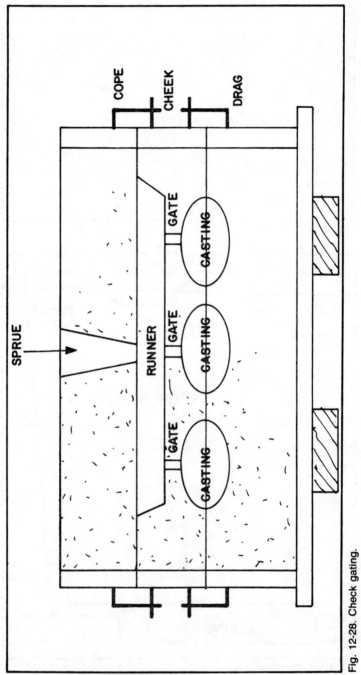

Fig. 12-28. Check gating.

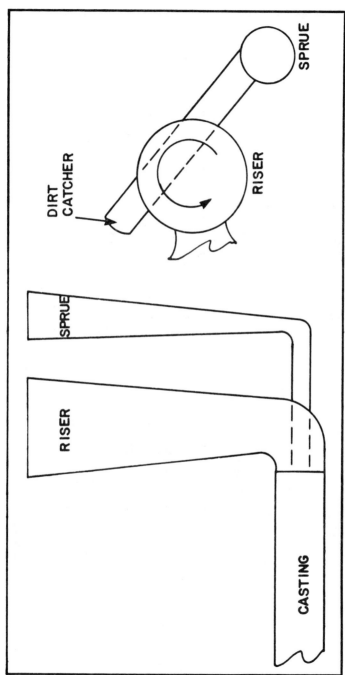

Fig. 12-29. Riser whirl gate.

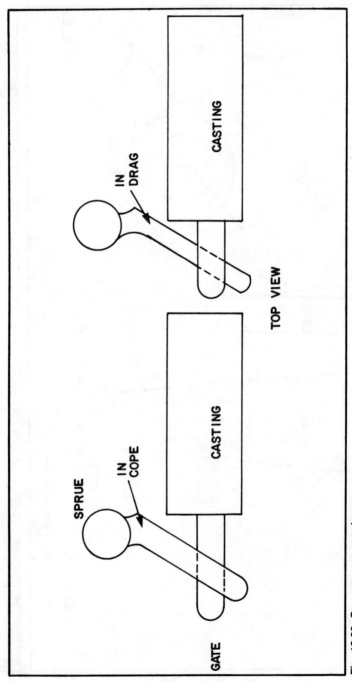

Fig. 12-30. Crossover-crossunderage.

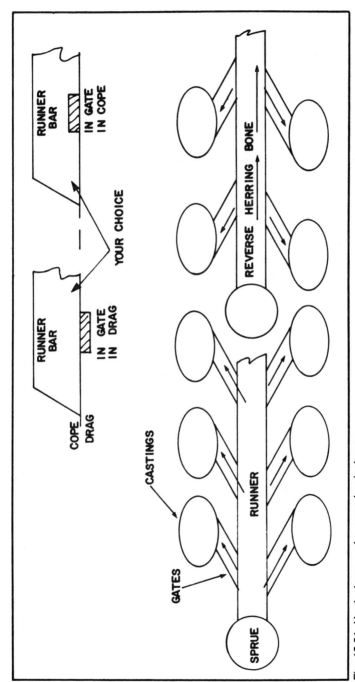

Fig. 12-31. Herringbone and reverse herringbone.

207

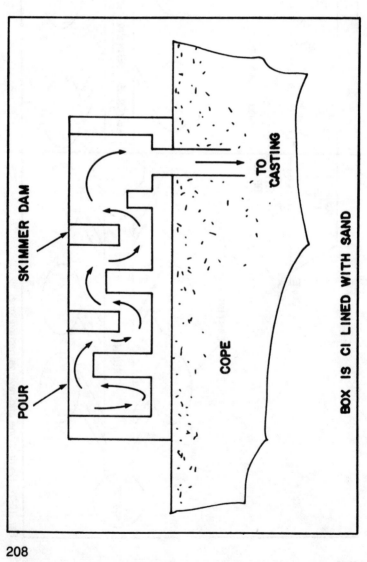

Fig. 12-32. Baffle-box gate.

POUR

SKIMMER DAM

TO CASTING

COPE

BOX IS CI LINED WITH SAND

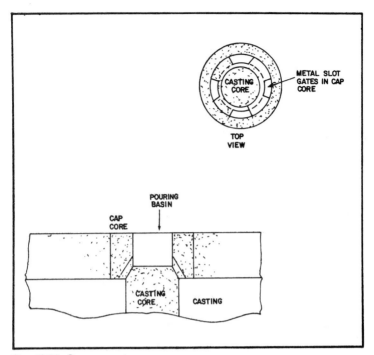

Fig. 12-33. Cap-core gate.

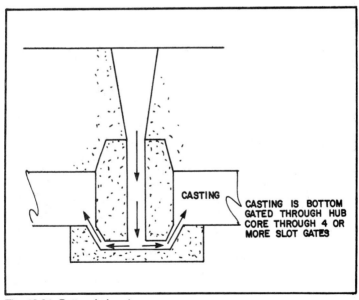

Fig. 12-34. Bottom-hub gate.

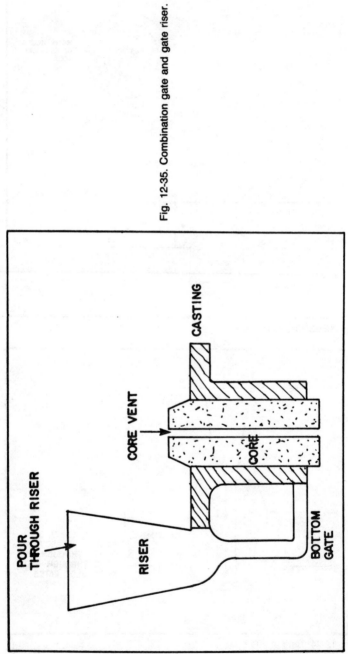

Fig. 12-35. Combination gate and gate riser.

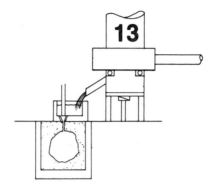

Revelations

If you are jobbing you are going to have the continuous problem with the people who have little or no knowledge of the alloys, their uses, and their specifications. The biggest problems come from the design engineers.

A TRAP

There are untold items out there today that are cast out of the wrong alloy simply due to lack of knowledge. Years ago in a large shop we were casting an item. The print called for manganese bronze with a minimum tensil strength of 95,000 PSI. The specifications called for 100 percent x ray elongation of 20 percent in 2 inches, a yield of 48,000 PSI etc., etc. As the part had no name other than bracket and its use could not be determined by what it looked like, it was assumed that the part was crucial due to the specifications.

The design of the part was not suited for casting in manganese bronze. Correct design is very important to success when you are casting manganese bronze. Well, to make it short the scrap total was very high; one bubble on the x ray and out the casting went (the price was high). The job became such a headache that—prior to coughing up the job—we decided to call the customer's engineer in and see if we could reach a solution such as a change in alloy, design, specifications or what have you. After scrapping untold hundreds of these castings for various reasons, we discovered that the casting was a bracket to hold up an ash tray for the pilot in an airplane. It could have been cast out of zinc (die cast) or plastic. It supported practically nothing, and was under no mechanical strain etc. Don't get caught in this trap.

ANOTHER TRAP

You wouldn't call for a high-strength tool steel shaft with a plus and minus tolerance of .001 of an inch for use as a wheelbarrow axle—now would you? The days when a fellow came into the foundry and said, "Look, this is what the item is going to be used for, how about making me a casting that will fit the bill?" are more or less gone. You are dealing with a new breed of high-technology boys with all kinds of fancy degrees and ideas. You must avoid traps. Mr. X comes in, is inflexible, and doesn't know what he is talking about in regards to an alloy, design, or specifications. You know it is not only a big problem from your end but could lead to bad feelings or even a lawsuit (product liability, more on this later). Don't take on the job. There are too many no-problem jobs out there. Don't do it regardless of how hungry you are. It's bad enough when you get involved in a job that *looked* like a piece of cake, only to wind up as a real bummer.

SAMPLE, SAMPLE

The XYZ machine shop has an order for 100 fairly simple-looking, no-problem castings. The company thinks they are simple duck soup. They look simple to you, but you have some reservations, some negative vibes, or your intuition is that there could be a hidden or unrecognized problem. Come right out and tell XYZ that, in order for you to give him a realistic price and delivery time, you would like to first cast a sample. Agree on a price for the sample, but make sure that it is understood the price is only for the sample. Additional castings could cost the company more or less per each item.

A PROOF CASTING

Often a proof casting is a must with a new pattern or design. One proof cast is used to determine if it's going to work out (machine, mate correctly with some other part, etc.) prior to simply jumping on it and find out later that it needs to be revised.

This proof casting will give you untold information. Aside from problems, you will get an accurate casting weight (weight with gates and risers, pouring weight), finished weight (degated, etc.), all costs, plus pattern problems. Then you can sit down with XYZ and come to terms whereby the company can get what they want and you can make a reasonable profit.

This system is not foolproof. I have seen cases where the proof casting was a winner and from there on out a losing proposition. That's part of the game.

TRY ONE FIRST

I once poured 55 flanged bushings—a simple casting, a simple pattern, and a simple molding job—only to have them all rejected for gas porosity that showed up when the customer machined them. Each bushing weighted 115 pounds cleaned up.

They were all poured from a large bull ladle that was green and not properly dried (wet ladle). Had we cast one and let the customer machine it first or simply took a rough cut through the bore ourselves, we would have caught the problem at that point. Oh yes, we recast and they (the second set) were perfect, but the profit (plus) had already gone up the stack.

Price It Right

A small simple casting would or should not need a trial run. If you have only one to cast and that's it, put a high-enough price on it to cover yourself.

Turn It Down

Turn down a complicated casting that you feel is beyond your ability, out of your line, too large for your capacity, or requires special rigging that you don't have. Let some other shop have it who specializes in the appropriate class of work.

I know of several foundries where anything under a ton would puzzle them and vice versa. Pick a class of work and weight that fits your shop and ability.

The Customer Shoots The Long Bow

Shooting the long bow is an old foundry expression for shooting the bull, not telling the truth or bending the truth somewhat. Most customers are prone to shooting (or wielding) the long bow. Some are quite apt at it and most convincing.

Example: Our long bow wielder shows up with a nice shiny pattern and starts off with his bow work. He has heard that you produce excellent work and thought he would throw some of his special work your way. He then goes into great detail of what a fine fellow he is, that his company is the finest, and—should you please him and his shop foreman—you are on the way to becoming rich. You look over the pattern he says has been used to make countless fine castings. Now, it doesn't take much smarts to tell if a pattern has been in the sand many times, a few times, once, or never.

I have often heard this story over and over for years and have been handed patterns that might have only one draw pike hole in it or none. Our boy might have a problem job coughed up by your competition. He could be price shooting or whatever. If he is on the up and up, he will simply hand you the pattern without a story and let you make the first move. If the same fellow shows up with a pattern, that appears to have been used many times, and starts off with some such tale, or that the AA foundry was not only sticking him, but did lousy work, you can bet he owes our friend AA some foundry money.

How do you handle this? Simple, get him to leave the pattern for you to figure him a price (do a little long bow work yourself). Say you will get back to him as soon as possible. If he leaves the pattern, and it will fit in a 12 × 16 flask or smaller, ram one up for a little molding test or have one of your athletes do it. Get on the phone whether he leaves the pattern or not and call the local competition. You will be surprised how cooperative they will be if the job is a bummer or the owner of this pattern is a deadbeat and owes them for castings or machine work. Sometimes the pattern itself is hot property.

Long Distance Long Bow Work

Long distance bending is quite common, it has been going on for some time, and it always will be with us. Anyway, what it amounts to is that they will load you down with work. The line is that profits are small but continuous; therefore your worries are over. They supply the patterns and in some cases the flask equipment. It sounds great. They set the price per piece, give you all you can do or want to do, but require a contract for at least a given amount (a run). At this point, they start talking in the thousands. Mr. Splendid is a casting broker and you are being sucked in. The contract is tight, the profit practically nil, and—if you run into problems—you are dead. It all sounds too good to be true. And it

214

is. Many firms hire or buy their castings through a broker, and his job is to get them cheap. Don't get involved.

It goes like this. You are sitting in the office wondering how you are going to keep afloat. The phone rings, you answer, and a soft feminine voice says, "Mr. Foundry, this is the Go Getum Mfg. Co., of New York, NY. Our Mr. Splendid, the president, would like to speak with you. One moment please." Now, Mr. Splendid doesn't jump on the phone at once—no way. He lets enough time go by so you can work up a mental vision of Mr. Splendid's firm.

By the time he does say "hello," or "Splendid here," you have already figured out that here's your angel. They have to be big, they are in the Big Apple, his secretary called, they didn't call collect, oh boy. After the appropriate wait, Mr. Splendid comes on the line. "Sorry to keep you waiting Mr. Foundry," He says that his firm wants to know if you have some open capacity to make some castings. If so, he can give you all you would like.

This is a common practice with some manufacturers of small tools, drill presses, wood lathes, and the like. They will split the work and job out the base to one shop, the spindle to another, etc. Nobody wins but them. I have seen more than one small shop go down the tubes this way. If you tie yourself up with cheap work, when the fat ones come along you cannot take advantage of them.

THE NON-FOUNDRY, FOUNDRY

In some cases, a fellow will set up an office and call himself a foundry. He has a bunch of prints printed up of stock items, fly wheels, bushings, gear blanks, etc. These prints have different names on them. For example, the fellow calls his firm Continental Foundry Inc., and he advertises that he will cast anything, any weight—brass, bronze, aluminum, steel, gray iron. Now, let's say you receive a print through the mail from the Monkey Mfg. Co., with a request to bid on casting 100, 500, 1000 and 50,000 of the item shown. You have two copies of the print, and they request that you mark up one print (add core prints, etc.) and return it to them with your bid filled out on the enclosed bid form.

So you do this and you never hear from these people. You figure you were too high or whatever. Actually, there is no Monkey Mfg. Co. at all. The Continental foundry is some joker in a 10- × -10-foot office sending out prints all over the place to every foundry in the country. If you were to send out 1000 prints and bid forms to that many foundries, asking each to bid on the same item, in no time you would know which foundries were hungry, which were fat, and which were begging for work. If you did this with a wide variety of items in a wide variety of metals, in short order you would have a good picture of what type of castings could be bought for and where.

Armed with this information, you find the fellow who needs the castings and you bid a price. Then you buy them from one of your file foundries (leaving a profit for you, the middle man).

You see these ads in the foundry magazines and trade papers: *Wanted: Foundry Capacity. Do you have any available capacity in cast iron or non-ferrous work. If so, contact the Monkey Co., we need castings in large quantities.* The ads are from fishermen, brokers, non-foundry foundries, and the likes. Avoid them.

"Free" Engineering

I was superintendent for a large jobbing foundry in the Midwest. Shortly after taking the job, I had a request from the boss patternmaker for additional help. He wanted a man to help him figure the bids and mark prints because he was swamped, and this took too much of his time away from overseeing of the actual pattern work. He received, from the front office, a huge pile of blueprints from firms all over the country every morning.

We were operating at about capacity and for approximately 20 customers who had been with us for years. Why was he going through all this bidding every day? He had several large files and countless boxes full of his copies of the prints and bids going back seven years. I found several firms that requested bids weekly, and had done so for years. Upon checking the records and the pattern storage, I found that we had never been successful in getting one job from these firms. Yet we continued to mark up prints and submit bids weekly.

Let's single one out. A radio manufacturing firm, which is still around, had been submitting blue prints and specifications for a period of seven years and had averaged four prints per week. In the seven-year period, we (the foundry) had never been successful in hanging one single job. The patternmaker should have gotten the message after the first six months.

I put on my gumshoes and decided to make a case of this one. They were only 50 miles north of us and, if they were buying castings at all, it would have to be somewhere within a 100 mile radius.

Let's shorten this tale. I found the foundry that was doing their work. Some 90 percent of this foundry's work was for this radio firm (for which we were continuously bidding). They were using us to do their engineering work for free and using the figures we quoted to keep the foundry doing the work inline. A neat trick. We broke up their little game.

This practice is quite common, but you can work it in reverse. You get a print to bid on from Co. A, you redraw the print, put a phoney name on it, and send it to your competitor. He marks it up, bids and returns it to your phoney mail drop. Then you compare it with what you would figure.

Of course this game can backfire. Most firms buying castings—even small lots, where the items are more or less stock, such as bushings and gear blanks—know what they weigh and approximately what they should cost them. Items such as a replacement casting (repair work), new items, machine parts, and prototypes are a different story.

Ferrous and Nonferrous

Often as not, a small gray-iron foundry decides to do a little brass and bronze castings along with their primary function, cast iron. Or a brass foundry decides to make a little gray iron. This rarely works out; before long you have all kinds of problems. If you maintain two separate sand heaps—one suited for cast iron and one suited for brass—in no time they both become contaminated beyond recovery.

It was common place in years past for an iron foundry to have what we called a doghouse foundry. This was a small brass foundry—usually consisting of a brass molder and a helper in a separate out building—with its own sand system, melting equipment, and core room just for nonferrous work.

This worked out fine. In some cases, the doghouse was several blocks or miles away from the iron foundry. The two men were skilled nonferrous founders and did it all—molded, melted, made cores.

Today it is common to find brass being cast in an iron foundry in gray-iron sand. The molds produced by gray-iron molders, using gray iron foundry practice, gating, and mold hardnesses, are all well suited to cast iron, but not for non-ferrous items. The product is a mess.

If there is a gray-iron foundry in your area that does this work, it's usually easy enough to talk them into letting you do their nonferrous work. The customer who buys, say 80 percent of his casting needs in cast iron and 20 percent in brass, usually thinks a foundry is a foundry and insists that the iron foundry do it all. If possible, they job out the brass work. The two do not mix.

ART CASTINGS

There are two classes of art castings in which the brass foundry can get involved. The first type is items produced in wax or clay by the artist sculptor. The second type of art casting work is classified as architectural bronze work. This type of work commands a good price and does not involve the artist. This work is usually designed by the architect; they are usually easy to work with and, being engineers, they will give and take in regard to the metal selection and pattern equipment.

This type of work covers a large variety of items such as small rail balls, plaques, grave markers, to large bronze doors. Specialty foundries, small brass foundries and bronze foundries can make a good living doing this class of work. Some of the castings are purely decorative while others can be both decorative and functional (doors, large hinges, door knobs, and countless other items).

If this class of work is the foundry's specialty, the largest portion of this work is usually polished, buffed, and—in many cases—patinaed or plated. This class of work requires precise foundry practice in regards to sharpness, detail, smoothness, and—above all—free from any porosity (due to gas or shrinkage) and any sand defects that cannot be polished out. Color matching is a real art here. Often the casting is to become a part of a combination of wrought brass or bronze (mill products), and the casting must match the mill product in color. This is a mean task.

ANTIQUE DEALERS

Antiques are a specialty field. Dealers are forever looking to match missing draw pulls, lamp parts, keys and all sorts of odds and ends. They are not quite as easy to deal with as architects. They all are price shooters and fancy themselves as real horse traders. You have to simply hold your ground and insist on your price. They will pay it if you don't back down. This class of work is usually done using an old casting as a pattern. Great skill is required in loose molding and in finishing the casting. The castings are usually lightweight, and 99 percent of the price you charge is for your skill and labor.

The dealers cannot shop around because these items are often unique and not a shelf item. Many that are extremely thin require hot metal, and perhaps polishing and or plating.

A great deal of this work requires bright nickel plate; some also require silver or gold plating. You must charge through the nose for this work or you will find it a losing deal. As to the plating, if it must be jobbed out, you do this and tack on 20 percent of the cost for your service. If it is your intent to specialize in restoration castings, you should have the polishing and plating facilities yourself.

ANTIQUE AUTO RESTORERS

For auto restorers, you will find that in a great many cases you are asked to replace an item of where there is none to use as a pattern. You have a drawing or picture and will have to make a pattern. With architects, artists, antique dealers, always get enough up-front money to cover yourself, plus a profit of some sort, should they blow out on you. Believe me they do.

INVENTORS

Now here is a fellow for you. All jobbing foundries, large and small, are plagued with inventors. If done right, you can make some good money but you must understand the game plan. Most inventors are sure that they have a winner, but 1 in 10,000,000 are successful. Most inventors, but not all, start off with a proposition that usually goes like this. "You make the pattern and castings for nothing and I'll cut you in for 50 percent, bla, bla." They are going to do you a favor and let you in on a real good deal.

Simply tell them that good deals keep you broke, and that the pattern is going to cost x dollars and the castings x dollars plus any machinework. Your policy is pay in advance with 50 percent now and the other 50 percent when he picks up his stuff. If he buys the deal, take his coconuts and jump on the job. If not send him to your competition. Inventors have busted more than one foundry.

SEMINARS

About a half hour after you start up your operation, you will start receiving in the mail, from every quarter, invitations to attend all sorts of seminars. These seminars cover every subject from sand casting, melting, how to run your office, OSHA, EPA, basic metallurgy, so forth and so on. They are mostly hot air.

Talk about long bow wielders. These fellows will, for a fat fee, drop some high-price pearls of wisdom on you. If you follow their advice, in no time flat you will be rich. The giveaway is easy to spot. Look at the credentials of the fellow giving the seminar and the subject. Subject: Modern Methods of Casting Cast Iron Rudders for Ducks. Presenter: Dr. A. Jake O'Toole, Ph.D. CCTD (Choo Choo Train Driver), etc.

You would be better off to find some old sand crab who has cast hundreds of duck rudders in gray iron over a period of 20 to 25 years. He will have had his hands in the sand and have bent his hairy ears on the fine points of duck rudder production. Old A. Jake O'Toole's bag is giving seminars. This is a business in itself, and not a bad racket at that. You pay your $300 to $400 or more up front. When you get there, they hand you, in book form, exactly what the fellow is going to enlighten you on using slides. You can take the book home and skip his long-winded deal or, better yet, simply go out and buy a good book on the subject.

We seem to be drowning in experts from every side. Television has an enormous amount of them each week. Each fellow has written a book and worked his or her way on a talk show to peddle the book. Child care written by a woman who has never had children. How to get rich with real estate by buying it with no money down. One fellow actually had a doctorate in leisure.

MONEY TO THE DUMP

When operating a small shop, in time you will have to cut the floor down to its original level (before you reach the ceiling with your head). This is especially true with a dirt-floor foundry. System sand will become unusable by virtue of too many fines, too much core sand, and dead and burned clay. You have to take this to the dump or start a mountain in the flask yard. If you run all this material through a riddle, you will be surprised as to the metal tools, castings, and other items you will recover.

Years ago there was a fellow who made a living cleaning up foundry floors for the metal he salvaged (along with pocket change, etc.). You can rollout a few bucks of goodies with every wheelbarrow you send to the yard or dump. Little items like this can bust a small operation.

SPEED VERSUS QUALITY

With a small foundry's percentage proposition, if you have one of a kind to cast it is imperative that you get it right the first time.

Molding speed comes with skill. I have seen molders who worked up a storm to produce an acre of dirty molds, producing a high percentage of scrap. When jobbing with loose patterns, carded patterns, patterns with follow boards, and original pieces used as patterns, this type of work requires great skill as a molder.

Take your time. It's not how many castings you produce per day, it's how many really fine, defect-free castings you produce. Great care and attention should be taken.

After the mold is closed, often a good clean mold to start with produces a sandy, dirty casting simply because the molder did not clean the top of the mold and cover the pouring basin with a little cardboard. Dirt can be introduced down the sprue, prior to pouring, and be washed into the mold cavity. I have seen molds spoiled simply from loose molding sand, falling unnoticed, from the molder's clothing (his hat or shirt pocket) and down the sprue when he placed the mold on the floor.

Dry sand from a dirty riddle going against the pattern, when riddling the facing against the pattern, will cause a dry spot in the cavity which will wash. The same will happen with pockets or streaks of dry sand in a pile of improperly conditioned and aerated facing or floor sand, resulting in a scrapper. Soft, uneven molding will cause no end of problems from burn in, swelling, washing, and drops.

When a mold is damaged during molding (such as a drop or damaged edge), unless this is very minor, shake the job out and redo it. Repairing a damaged green sand mold properly requires great skill. Even if the molder can do a successful repair job, it usually requires less time to shake the flask out and start over. To spend more time in trying to effect a satisfactory repair then to make a new mold isn't smart.

If your flask equipment is rickety and the pins and guides are not accurate and in good shape, you are destined to a great deal of failure.

When molding loose work in a small snap or tight flask, always, after closing the mold, reopen it and blow it out again. For loose, hand rammed molding, I have always been in favor of using the hand bellows for blowing out the sprue, the gating system, and the cavity (but not the compressed air hose). The bellows will do far less damage due to the control and gentle puffing action.

If you dust the mold cavity and gating system lightly with fine wheat flour from a parting bag prior to closing, you will find that you will not only produce cleaner castings, but the color of the castings will be like they should. They will not be mottled with patina spots or dark, off-color skin.

This is an old practice; at one time no molder worth his salt would be without a flour bag. You always pop the bag first away from the mold prior to dusting the mold to get rid of any dirt or loose sand stuck to the bag or picked up from its resting place. This seemingly minor move of popping the wheat bag clear of the mold to be dusted is one of the little details in producing a good casting.

The wheat flour, when moist, forms an adhesive (wheat paste). When shaken on the damp surface of the sand mold, it soon forms a thin skin that will hold together and prevent washing of the sand grains from the surface and into or on the casting. A second function is that wheat flour, when heated, has a strong reducing power, preventing the incoming metal from oxidizing and leaving the surface of the casting bright and free from oxides.

The atmosphere in the mold cavity is reduced at once to a reducing atmosphere in place of an oxidizing atmosphere when no wheat flour is used. This effect is also produced by the addition of fine sea coal or wood flour to the facing sand. When the sea coal and or wood flour is heated by the incoming metal, it is converted to CO then CO_2. I have worked in shops where, just prior to pouring, the molds were filled with CO (carbon dioxide) from a tank with a lance that was introduced into the sprue.

It only takes one test to make a believer out of you. Pour one red brass mold as is and one with a flour dusting or with a facing containing sea coal or wood flour. There is no comparison in the appearance of the castings. You might ask if this is so then why has it been more or less abandoned and no longer mentioned in foundry texts relating to practice.

The answer is quite simple. As we rush forward in the pursuit of more and faster production, dusting a brass mold with wheat flour was deemed by some time and motion "expert" as a waste of time and effort and without merit. Every so often, much to my delight, I will come across an item referring to a newly discovered sand additive or some new practice. This is simply something we knew and practiced years ago—forgotten low technology. These fellows reinvent the wheel quite often.

NONFERROUS MOLDING
SAND WITH SEA COAL ADDITIONS

While I am on the subject of additives, let's talk about sea coal and brass and. Sea coal is finely ground bituminous coal used in sand mixtures to prevent fusing of the sand to the castings. This is the usual definition given in various foundry manuals. There is more to it then meets the eye. Yes, it is refractory, so this would help to prevent burn-in.

Sea coal is considered only as an additive to sand used for cast-iron work. Never is it mentioned as being used in brass or bronze facing sand or system sand. It is widely used in iron work and sold in various grades. The only grade we are interested in here is what is known as *silk bolted* or *air-float* sea coal. It is extremely fine coal that looks like a coal flour.

A good specification for brass work would be air float sea coal: volatile matter 38.5 percent, fixed carbon 53.8 percent, ash 4.81 percent, sulfur .9 percent maximum, moisture 2.5 percent (what we used to call grade "E"). It must be high in volatile matter to ensure quick and even ignition. This ensures that the mold cavity is quickly converted from an oxidizing atmosphere to a reducing one, and remains so until the casting has solidified. My preference is a sea coal that is fine enough that 98 percent of it will pass through a 200-mesh screen.

Let's look at a typical gray-iron facing sand for very light work: 100 pounds sand, 5 pounds bentonite, 5 pounds fine sea coal. In this formula you have 4.5 percent sea coal. Now this would be a bit high in sea coal for brass, other than heavy chunky castings, but you would be surprised as to how fine of a finish and color a brass casting would be in this sand. *The Complete Handbook of Sand Casting*, TAB book No. 1043, describes rebound facing. The formula is 3 parts airfloat sea coal, 1 part goulac, and 10 parts natural bonded molding sand.

You will find that the castings, when faced with this mixture, will have a good color, sharp detail, and peal easily (the sand will fall away from the casting). The little time it takes to face the pattern is well worth the effort.

Most formulas for sands using sea coal run from 4 percent to as high as 6 percent sea coal for brass and bronze. I wouldn't go over 3 percent sea coal for general work. It is useless in aluminum sands because the aluminum is poured at such a low temperature not much happens to the sea coal other than a dirty casting.

The sea coal should be made of high-quality coal and not what is known as slack coal (high in slate which is considered an impurity). You can replace sea coal in molding sands and facing sands with charcoal flour, but this is usually too expensive. When printing back a pattern to produce an extra-smooth face of great detail, charcoal can be used as a facing parting.

Let me explain. Print back is where a pattern is removed from the sand and the face of the cavity is dusted with graphite or high early portland cement. Then the pattern is returned to the cavity and rapped down to set the dusted-on facing tightly against the sand. This presents a smooth surface in the mold cavity and this is where the charcoal comes in. The pattern is rapped and lifted out, and the cavity is dusted with cement or graphite from a dusting bag (old sock). Then to prevent the cement or graphite from sticking to the pattern when it is returned to the cavity, it is rapped and again drawn out, and a light coating of charcoal is dusted on top of our cement or graphite coat. Then it will be driven into the sand and will not stick to the pattern when the pattern is withdrawn again.

Sea coal can be used here instead of charcoal with the same results. See Fig. 13-1.

THE SUPER PRECISION CASTING

There has been a continuous industry move toward precision casting in an attempt to eliminate machining or other finishing. Many parts that were sand cast and machined have moved to precision investment casting. In a great majority of cases (not all), the only thing accomplished is that the cost per unit has gone up. There has been a lapse of memory or again a loss of technology. Always consider carefully the many basic machining methods available to

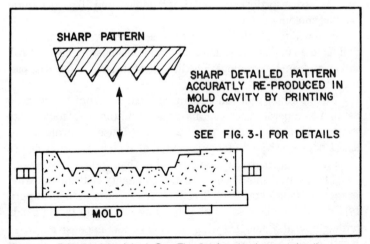

Fig. 13-1. Print back revisited. See Fig. 3-1 for step-by-step details.

machine easily and at a minimum of cost before you attempt to try to save machining.

Even with basic simple tooling, it is much less expensive and more accurate to machine a slot and drill a hole than to try to cast it to a close tolerance. You must remember that the technology of machining has also advanced in leaps and bounds. Most brass and bronze is easily and quickly machined. If your melting practice is correct and your house is clean, you can recover your borings and turnings.

CLEAN FOUNDRY

Little is ever said about housekeeping in the foundry other than from the safety standpoint. Let's look at it from the angle of producing better brass and bronze castings. This subject might seem completely out of place in a book on bronze and brass casting, but not so. It is assumed by most foundry writers that you know that a dirty shop produces dirty work. I have yet to read one article on foundry practice in reference to the effect a dirty shop has on the quality of castings.

I once took over a medium-size brass foundry that produced valves, and it was in the red. My job was to put it into the black, and when I did so I would receive 10 percent of the profit as a bonus each quarter. Let's look at some of the problems and solutions. Not all of these problems were found in this one shop. Nevertheless, they all relate.

Note: With the advent of OSHA and EPA, and their quest to safeguard workers from illness and harm, they came down hard on dirty operations. Much to the surprise of many an owner forced to clean up his operation, remarkable improvements were obtained in the casting quality.

A molder who wears clean clothes, and does not let himself become covered with sand, usually produces a better casting than his opposite. You cannot make clean work while molding with dirt. A heap sand that is contaminated with shot metal, core sand, core butts, core wires, gaggers, cat droppings, clay balls, cigarette butts, gum wrappers, and match sticks will only lead to scrap castings. Many a casting blow goes unexplained, but was caused by a clay ball, cigarette butts, or cat croppings that wound up on or near the cavity surface during molding (and simply blew up when heated by the incoming metal causing a blow defect). See Fig. 13-2.

Many a drop before closing or (worse yet) after closing has caused a complete scrapper due to match sticks, dry unbonded core

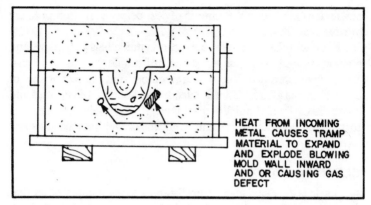

HEAT FROM INCOMING METAL CAUSES TRAMP MATERIAL TO EXPAND AND EXPLODE BLOWING MOLD WALL INWARD AND OR CAUSING GAS DEFECT

Fig. 13-2. Tramp material in molding sand is very often the cause of a scrap casting.

sand, a bit of paper or what have you forming a parting causing a drop. See Fig. 13-3.

A core butt that has sucked up a large amount of moisture, when rammed up too close to the pattern, will blow or produce enough steam to badly gas the casting.

Let's talk about dirty ladles and crucibles and improper skimming. Aside from nonmetallics such as particles of refractory material getting into the casting, a dirty crucible or ladle with an oxide ring buildup (when filled above this ring) sometimes results in a portion of the dirty oxide becoming dislodged and then poured into the casting. This fragment introduced into mold, which is being

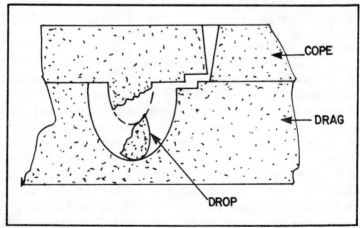

Fig. 13-3. Drops caused by random dry sand or tramp material.

226

filled with deoxidized metal (some of the residue phosphorus, lithium or other deoxidizer) will react with our oxide swimmer, deoxidizing it. As it gives off its oxygen, it will be propelled along until it is reduced, leaving a trail of progressively diminishing "oxygen" bubbles behind it. This shows up in the casting as voids, and is known as a duck flught. The only cause is deoxidizing oxide swimmers. Most brass and bronze founders are puzzled as to what causes them and why. See Fig. 13-4.

Duck flights are also introduced into the mold by improper and or incomplete skimming. The proper method is to use a clean oxide, slag-free skimmer. The metal in the ladle or crucible is deoxidized, if required, and then a small handful of clean, dry silica sand is thrown on top. This chills the slag and makes it an easy task to collect completely as a nice ball (like taffy). The skimmer removes 98 percent of the slag and oxides that were not reduced. Of course, a bottom pour ladle or crucible also helps you here.

For the proper pouring, pour close to the sprue as possible, keeping the sprue choked throughout the pour. You must establish the choke with the first splash of metal from the ladle.

Skimmer gates and skimmer cores help, but with or without these aids the first requisite is clean ladles and crucibles. Ladle linings should not be cracked or broken and red hot.

Dirty or rusty chaplets will blow and not knit. Dirty or rusty chills will kick and blow, causing defects at their point of contact.

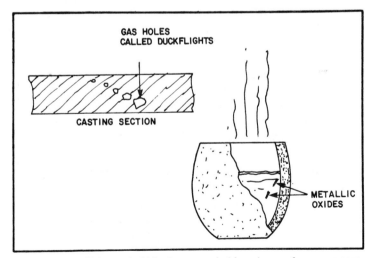

Fig. 13-4. Duck flight gas bubbles have puzzled foundrymen for many years. This is a simple defect to prevent.

A dirty molding shovel will cost you in molds per day. It takes a lot more energy to shovel with a dirty shovel caked with molding sand and rust. Dry, caked sand dislodged from a dirty shovel will cause partings, washes, and drops. Your shovel should be cleaned often and coated with bayberry wax. Never leave it dirty at the end of the day and place it across your molding bench.

Do not leave the shovel stuck in the floor sand or facing heap to rust overnight. A dirty melt area, oily sprues, and generally poor housekeeping of the furnaces and area is the cause of untold casting defects. I could carry on for pages, covering each problem in detail, but I am sure by now you are aware of the dirt problem.

This also goes for dirty encrusted riddles and riddles that have holes or tears in them. The ceiling joists and rafters that are never cleaned, raining down filth from above into the sprues, open molds, dirty and rusty flasks—you name it. It all adds up to scrap.

VARIOUS LIES

- [ ] Our company will make you rich.
- [ ] Do it cheap and we will give you volume.
- [ ] This casting is duck soup.
- [ ] Give us a price for thousands.
- [ ] Your competitor will give us a better price. Can you beat it?
- [ ] We like you.
- [ ] Cast my artwork and when I sell it I'll pay you.
- [ ] The check is in the mail.
- [ ] The man who signs the checks isn't here now.
- [ ] Leave the castings and we will send a check in the mail.
- [ ] The last foundry did better work.
- [ ] We pay only by the pound for castings.
- [ ] The Porsche is paid for.
- [ ] The last foundry didn't have any trouble with this job.
- [ ] We are doing you a favor letting you cast for us.

The revelations just given in this chapter represent a small portion of potential problems. A careful study of them will pay off in big dividends. It's the little things that break both small and large foundries, and most of these little traps or pitfalls have nothing to do with your ability to produce good work.

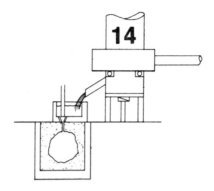

The Business

Foundrymen are notoriously poor businessmen. Unless you are operating a foundry for your own amazement, you are looking for a profit. I gave the information found in this chapter as a lecture to a group of businessmen several years ago. One fellow—who has a Ph.D. in business administration—told me, "Mr. Ammen, I have learned more about business in listening to you for one hour than all my years in college." So pay attention!

THE TUBE

Let's imagine that the tube shown in Fig. 14-1 represents the ABC Brass Foundry. The inside diameter of this tube is just large enough to roll a coconut through. The length of the tube represents the size of the ABC Foundry. Therefore, the longer the tube the bigger the business, the shorter the tube the smaller the business.

Now let's collect some coconuts and imagine that they have a value of $100 each. The value per coconut represents money (earned money, capital). After all, money was created to have a flexible barter media. It was supposed to represent gross national product or some fellow cutting wood, mining coal, casting brass, etc. You noticed I said *was*.

We now have the tube and a quantity of coconuts. See Fig. 14-2. Let's suppose that we have started up the ABC Brass Foundry of X size, and we proceed to put the coconuts in the tube. Before we get the tube full, we run out of coconuts. See Fig. 14-3.

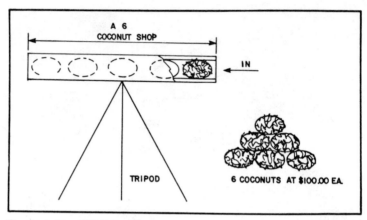

Fig. 14-1. The ABC Brass Foundry.

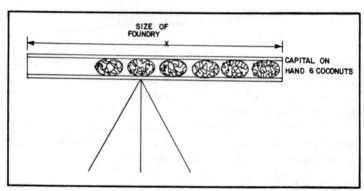

Fig. 14-2. The X foundry requires eight coconuts.

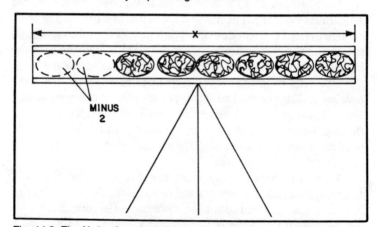

Fig. 14-3. The X-size foundry is too big for the available coconuts.

It becomes apparent that we are short of capital. In other words, we simply don't have the coconuts for a business this size. It is necessary that the tube be full all the time so that when we put in a coconut one comes out the other end. Now we see the longer the tube the more money (coconuts) we must have (which represents capitalization, fixed overhead, etc.).

We could shorten the pipe, reduce the size of the ABC Foundry, or go to the bank and borrow some coconuts from our friend the banker (not an easy task). Suppose we put up a good song and dance and hit a winner. Now we have the bank's coconuts in the tube along with ours.

Let's see how this is going to work out. We make a casting for a local machine shop. That costs us one coconut, and we trade the casting to the machine shop for two coconuts. We push them into the pipe and two come out the other end; we have a net gain of one coconut.

This is one of the banker's coconuts, however, and only really 80 percent of a coconut because he charges us 20 percent of each coconut (which he hopes to recover and shove into the bank's tube). It is obvious that, in order to make a profit, we must come up with less expensive coconuts to put in than the coconuts that come out. Our first casting cost us one coconut ($100) and it gave us back two coconuts: one $100 coconut and one $80 coconut (the bank gets $20). Did we net one coconut or 80 percent of one coconut?

Now the plot thickens. The next casting cost us more due to a mistake; we had to cast it twice instead of once. We had an additional cost of extra labor, lost metal, melting loss, etc.

This casting has a barter power of only two coconuts. It cost us 1 1/2 coconuts so we have a net gain of only 1/2 of a coconut. The next casting, with a two coconut barter power, is going to have to be produced for the cost of half of a coconut. *Note:* If we pay a worker two coconuts to produce one coconut of barter power, someday, somehow some poor fellow is going to have to produce two coconuts of barter power for one coconut to even the score.

SINKING FUND

Say we have problems (unforeseen costs or a strike) and we run out of coconuts to shove through the pipe. Everything comes to a halt. When the banker comes for his coconuts, we say tough. We have had it. If we would have, at the beginning, held aside some coconuts in reserve we could possibly have made it past the problems. This is where a sinking fund comes in.

231

When you are putting in $100 coconuts that cost $25 or $50, and pushing out coconuts worth $100 each, you should skim off 10 percent minimum (or 10 coconuts per 100 coconuts) and set these aside to use when you hit a point where you are putting coconuts in at $150 each and getting out coconuts with a barter power of less than $150 each.

If you have a 1000-coconut pipe and 100 coconuts belong to the bank, you have a problem. If 900 of the bank's coconuts are in your business and 100 are yours, the bank has the problem.

SHORTEN THE TUBE

Let's say that, after a period of operation, you decide that the tube is simply too long for comfort or to realize a profit. If the tube were shorter, you feel that fewer coconuts would be needed to keep afloat and make a few coconuts. What do you do? It's simple if only your coconuts are involved. You owe nothing so you simply saw the tube off where you feel is best and hang in there.

Say the bank's coconuts are in the tube, and you bought a big, new air compressor from the local compressor company on the time payment plan. In reality, you have the bank's coconuts and the compressor company's coconuts in the tube. Where do you cut? Or can you cut the tube? You have little or no choice because you are in business with these fellows. No way are they going to let you cut off their coconuts.

TIP AND RUN

You could tip the tube, let the coconuts run out the end, grab them, and run to points South and hope they never catch up with you.

SBA'S COCONUTS

Say you qualify for a nice fat SBA loan for 90 percent or better of coconuts required. Then you have Uncle Sam's coconuts in the tube. If things go great, it could work out for you. If not it's a walk away. Don't run from this one.

BRAZIL

I just watched a TV news report. This program went into great detail regarding the huge debt the Brazilian government owes the various banks around the world. The problem is that they can't even pay the interest much less anything on the principal. We are talking one huge amount of coconuts. The program went on and on as to why and what to do, plus how did the banks get into this fix.

It is actually very simple. The lenders simply got greedy and now have their coconuts hung up in Brazil's tube. Brazil wants to and has to cut back by cutting the tube (shorten up, pull in the old horns, etc.). The lenders have to put up more coconuts in an effort to save their coconuts. It might not work out anyway. Good after bad. If they lose, do you know whose coconuts they are going after to save themselves. You bet, the taxpayer's coconuts. It is going to be interesting to see how this one works out.

INCORPORATED

A corporation is defined as a fictitious entity into which the legislature has breathed life. It has some advantages and some disadvantages. The decision to incorporate your tube should be carefully studied prior to jumping in. Usually when someone asks advice, they simply are looking for someone to agree with them. Make your own decisions.

PUT AND TAKE

In the coconut game, the bookkeeping is as simple as the tube and coconuts. It's actually the same game. You have a large ledger with two columns (one marked Put and one marked Take). The book is titled *Put & Take Ledger*. You simply record each time you put in and each time you take out coconuts. These two figures, plus your checkbook, will tell you exactly whether you are making money, losing money, gaining ground or losing ground. Don't laugh at this.

A TRUE STORY

Years ago Sam J. Pitre, Jr., and yours truly had a foundry and pattern shop in New Orleans, Louisiana. We were in business together, but not in business together. It was Ammen's foundry and Pitre's pattern shop. We hired a bookkeeper. He proceeded to set up a set of very complicated books (debit, credit, etc.), and before he got it all together he was drafted. We got another fellow who informed us that the books the first guy set up were not only wrong but not the system he knew.

Start all over; we scrapped $150 worth of books. The new fellow started his foolproof system consisting of quite a few more books than the last guy. Because neither Sam nor I had any experience in the racket, we gave him full rein.

All the time this fellow and the first guy were doing their thing, we were busy making castings and patterns and trading them for coconuts for the tube. From the day we opened the doors, Sam kept the big Put and Take book and our checkbook. Well we didn't care what the bookkeeper said. We knew whether we put in more coconuts than we took or took more than we put, and if we were making or losing money or swapping coconuts.

Well, in time we decided to close down and move on, and because the business was a corporation we had to file a bunch of papers and prove a bunch of nonsense. To make a long story short, the bookkeepers books were useless in resolving any part of the business for the tax people. We took along our put-and-take book and the checkbook. That did the trick. Without the put-and-take book, we probably would have been in jail or still trying to untangle the business.

PAID HELP

If you have to hire help (molders, etc.) for each 100 coconuts you have to lay out to a worker in take-home coconuts the worker must produce at least 3 hundred coconuts of barter power for you. If he doesn't produce enough, you must raise the price of the castings to the customer to make things come out at least one for the fellow and three for the pipe.

This way you can price yourself out of the game very quickly and find your customer buying overseas where one coconut is paid out for 8 coconuts produced. This has happened to large segments of our industry. This is the long pipe game. If you keep the pipe short enough and make small job lots—one of this and a few of that for which you can charge a premium—the chance of your customer going to Japan or Taiwan is very remote.

THE COMPLICATED TUBE

You know what can happen if the tube is too long or you allow it to get too long. You can start out with or work into a complicated multitube. This is where people get very greedy or very foolish.

Let's say you are doing fine with a straight tube and enough coconuts moving (making money but not growing). You spy some other business that looks good, but it is not really your bag. If you are smart enough to do what you do, why not this? Like the fellow who says I think I will become a doctor even if it takes a week.

Well you add this to your primary tube, and because it is not the same type of business you cannot put it in line with what you have, you try to attach it somewhere into the system. See Fig. 14-4.

Soon you find out that, for this new business attached on to our ABC foundry business, the tube must be larger and take bigger coconuts to make it go. You thought you hooked into the prime tube at the right spot (Fig. 14-3) so you could take advantage of both divisions of the business. But this is not the case. You are going to have to do some cutting, adding, and rearranging.

You could go several ways. Figure 14-4 shows that you have enlarged the section X of your original business to accommodate the coconuts from ABC foundry operation and those of the new addition (the ABC FlyPaper Co., a division of the ABC Foundry Co.).

Figure 14-4 shows that you have two different size coconuts meeting at the junction and trying to go through section X. This section has to be large enough to accommodate the large coconuts. Let's look at what can happen.

There is a great number of ways things can jam; I'll show two. These log jambs stop the flow, and the more you push the worse it gets. See Fig. 14-5. Should you have done this? See Fig. 14-6.

You could move the ABC flypaper division across town, out of town, or better yet sell it off. Of course, if you elect to keep it but move it, you know that you are going to have a lot of duplication (two offices, etc.).

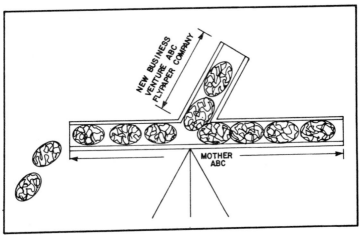

Fig. 14-4. The ABC Foundry branches out?

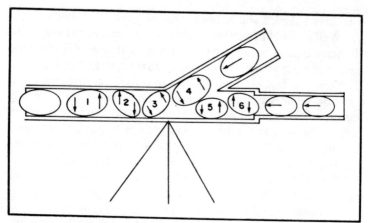

Fig. 14-5. A log jamb locks everything.

THE BALANCE POINT

In order to function, the tube must balance. Like an equation, both sides must be equal. If we add to one side we must add to the other. If we cut from one end we must cut from the other.

A true story. Some years ago I was the superintendent of a large steel-and iron jobbing foundry in the Midwest. This old, established shop made quality stuff from a few ounces to castings a ton or more. The tube resembled Fig. 14-7.

We had a good balance between the production end—the business, molders, melters, etc—and the nonproductive end (bookkeeping, an endeavor necessary to the business but it is a direct cost and nonproductive). The nonproductive end had to be paid for out of surplus coconuts from production.

Expansion. Expansion is rather simple if you have the

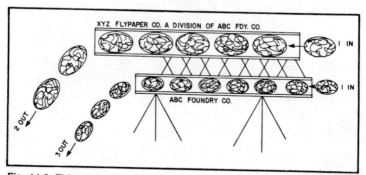

Fig. 14-6. This new business arrangement might work.

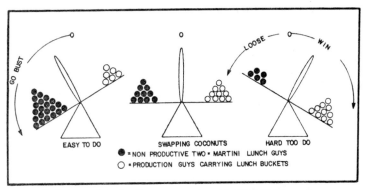

Fig. 14-7. Balanced production versus nonproduction costs.

coconuts for it or can borrow the coconuts. You use these to expand your production, update your machinery, and to add melt capacity. In doing so, you probably will have to add more to the office force (a few more non-productive coconuts can be tolerated). Don't get this confused with dead coconuts. If you do this carefully, you can maintain the delicate balance of the tube.

Reverse Expansion. What the firm did was to imagine or think that payroll was production or of a productive nature, and that the major expansion must be done in this direction. They forgot the whole concept that money was a flexible barter and had to represent production (making castings, etc.). They imagined that money represented money.

This concept is present today with drugs, food, tobacco, etc. Each time an item receives a warning label or a chemical additive is added to a food product that could be harmful or be carcinogenic, we will finally reach the point that everything is carcinogenic and/or poisonous. If everything is poisonous then nothing is poisonous. Living becomes carcinogenic so nothing is carcinogenic.

So these fellows reached the point where money was castings and castings were money. Then money was money so nothing was money.

Once they reached this point of thinking, they started to expand. They added on several efficiency experts, some time-and-study people, some time-and-motion people, lots of sales engineers, bookkeepers, purchasing agents, and several sand technicians. They expanded the lab and built a new plush office. They also reached a point were coconuts equalled coconuts in the equation. It looked something like Fig. 14-8.

They also tried to convince those of us out in the shop that all

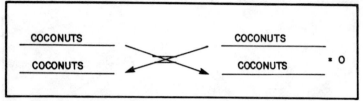

Fig. 14-8. The coconut equation.

that black stuff floating around in the shop air we were breathing was good for us. We did finally reach the point where the contamination was so high that nothing was contaminated. Oh yes, the tube looked like Fig. 14-9. It tipped over like an upside down pile of cannon balls. The old pyramid game. See Fig. 14-10.

By now I am sure you have the drift of the coconut-and-tube theory. With no more help from me, you could expand it to great lengths.

Uncle Sam. Don't look to Uncle Sam for help, Uncle Sam simply doesn't have the coconuts. This option went out when we got off of the gold standard and printed notes as needed.

Deficit. The government can operate at a deficit (for how long?), but there is no way you can operate a brass foundry for long at a deficit. If you can take from your put-and-take book without putting, you soon run to the point where there is nothing to take.

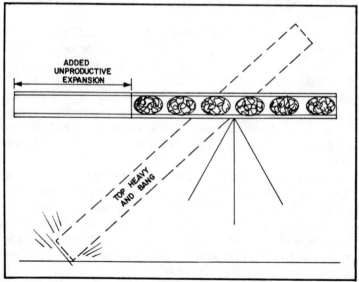

Fig. 14-9. Get off balance and go bust.

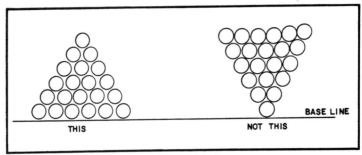

Fig. 14-10. Build your foundry from a broad base.

Pinto Beans and Brass Castings. I believe that the ultimate currency will some day become pinto beans and brass castings—objects you can barter with. The beans have an advantage; you can eat them. Brass is tough eating so swap the brass for the beans. I expect readers to expand on the coconut and tube idea.

You can go on and on. There is much talk in the news media about the trade deficit we have with Japan. Along with charts, graphs, and high-technology explanations and theories as to why and exactly what it is or means, there is very simply a put-and-take book entry. We purchase more from Japan then they do from us. Now that's simple and very easily equated in our put-and-take bookkeeping system. Coconuts in the tube versus coconuts out of the tube. Think about this carefully.

Appendix:

Sand Mixes

Half & Half:

Brass sand for general all around work.
50 percent Albany 00 or similar.
50 percent 120 mesh sharp silica.
4.5 percent Southern bentonite.
4.5 percent Moisture
2 percent Wood flour 200 mesh.

72 Hour Cement Sand:

Mold first day, wash second day, pour third day.
10 parts silica sand.
1 part Hi-Early Cement.
1 percent by weight 200 mesh wood flour.
Oil patterns to obtain a parting when molding.
On the 72 hour cement sand the reference to wash the second day refers to a core wash. This type of mold would be considered a core sand mold.

Synthetic Mix for Light Brass

160 mesh wash float penn silica
2 percent wood flour

4.5 percent southern bentonite'
4.5 percent moisture

Synthetic Mix for Heavy & Medium Brass

Nevada 120 silica
2% 200 mesh wood flour

4.5% southern bentonite
4.5% moisture

Rebound Facing:

3 parts air float sea coal
1 part Goulac
10 parts molding sand.

Synthetic Mix For Brass & Bronze

AFS fineness 60 to 80
95.5 percent by weight dry sharp silica
4.0 percent 50/50 Southern-Western Bentonite

(98.5 percent silica content)
0.5 percent corn flour
1.5 percent Dextrine

Dry Sand Mix for Brass & Bronze

Use system sand tempered with glutrin water, 1 pint of glutrin to 5 gallons of water.
Bake till dry at 350°F (5 percent pitch can be added to the facing sand).

Cement Sand Mix for Large Brass & Bronze

6.5 percent moisture   12 percent Portland cement.
81.5 percent sharp silica sand approximately AFS 40 fineness.

Facing for Plaques & Art Work

5 shovels fine natural bonded sand (Albany or equal).
¼ shovel powdered sulphur   ¼ shovel iron oxide.

Spray mold face with molasses water no later than 10 seconds after drawing pattern (this is one of the most important steps in producing a plaque). The molasses water in this case consists of one part blackstrap molasses to 15 parts of water by volume. Allow the mold to air dry a bit before pouring. Brass sand can be made up synthetically or partly synthetic and partly natural sand by choosing the correct base sand for the class of work and bonding with 50:50 south and west bentonite. A very popular all around mix is the brass mix.

Brass Mix:

90 pounds AFS-90-140 grain silica
10 pounds naturally bonded sand 150-200 fine.
4 pound southern bentonite
1 pound wood flour.
    To make any nonferrous mold suitable for skin dry work or dry sand work, simply temper facing with glutrin water or add up to 1 percent pitch to the facing.

Hi Nickel Facing Sand:

15 parts silica AFS G. F. 120 and one-half part bentonite.
3 parts system sand and one-half part fire clay.
    For dry sand molds a good all around sand for brass and bronze is a brass and bronze mix.

Brass & Bronze Mix:

Dry new sand, 95.5 pounds
Bentonite, 3 pounds
Corn flour, 0.2 pounds
Dextrine, 1.3 pounds
    In general for aluminum you can use a much finer base sand, synthetic or naturally bonded sand, due to the low melting and pouring range of aluminum. As a guide line, the two sands given here would cover 90 percent of all aluminum casting work.

Brass Fluxes:

● 100 pound dehydrated borax
  77 pound whitting
  50 pound sodium sulphate
● 50 percent Razorite
  50 percent Soda ash
● 8 parts flint glass
  1 part calcined borax
  2 parts fine charcoal
● 5 parts salt
  5 parts sea coal

  15 parts sharp sand
  20 parts bone ash
● 25 percent Soda ash
  25 percent plaster of Paris
  25 percent fine charcoal
  25 percent salt
● 50 pound glass
  55 pound Razorite
  5 pound lime

Brazing Flux:

50 percent boric acid
50 percent sodium carbonate.

Hard Bearing Bronze

Manganese—4 percent
Fe—4 percent
Al—7percent

Commercial Yellow Bronze

Sn—1.5 percent
Zn—26.5 percent

Statuary Mix #1

Zn—3 percent
8 ounces of Pb per
100 pounds

Statuary Mix #2

Cu—90 percent
Sn—5 percent
Zn—5 percent
8 ounces of Pb per
100 pounds

Statuary Mix #3

Cu—90 percent
Zn—7.5 percent
Sn—2.5 percent
Also used as a tablet alloy

Statuary Mix #4

Cu—88 percent
Sn—6 percent
Zn—3.5 percent
Pb—2.5 percent

Commercial Brass

Cu—64-68 percent
Zn—32-34 percent
Fe—2 percent max
Pb—3 percent max

Gun Bronze

Cu—87-89 percent
Sn—9-11 percent
Zn—1-3 percent
Fe—0.06 percent maximum
Pb—0.30 percent maximum

Ornamental Bronze

Cu—83 percent
Pb—4 percent
Sn—2 percent
Zn—11 percent

Red Ingot

Cu—85 percent
Pb—5 percent
Sn—5 percent
Zn—5 percent

Pressure Metal

Cu—83 percent
Pb—7 percent
Sn—7 percent
Zn—3 percent

Gear Bronze

Cu—87.5 percent
Pb—1.5 percent
Sn—9.5 percent
Zn—1.5 percent

Bronze

Cu—89.75 percent
Sn—10 percent
Ph—0.25 percent

88-4 Bronze Ingot

Cu—88 percent
Sn—8 percent
Zn—4 percent

Bearing Bronze

Cu—80 percent
Pb—10 percent
Zn—10 percent

Lead Lube

Cu—70 percent
Pb—25 percent
Sn—5 percent

Manganese Bronze

Cu—60 percent
Zn—42 percent
Balance temper depending
on grade desired

Nickel Silver

Cu—61 percent
Zn—20 percent
Ni—18 percent
Fe—1 percent

Tough Free Bending Metal

Cu—84.5 percent
Zn—10 percent
Pb—3 percent
Sn—2.5 percent

Bell Mixes

(White) Table Bells
Sn—97 percent
Cu—2.5 percent
Bi—0.5 percent

Swiss Clock Bells
Cu—75 percent
Sn—25 percent

Silver Bells
Cu—40 percent
Sn—60 percent

Best Tone
Cu—78 percent
Sn—22 percent

House Bells
Cu—78 percent
Sn—20 percent
Yellow Brass—2 percent

Sleigh Bells
Cu—40 percent
Sn—60 percent

High Grade Table Bell
Sn—19 percent
Ni—80 percent
Pt—1 percent

Special Clock Bells
Cu—80 percent
Sn—20 percent

Special Silver Bells
Cu—50 percent
Zn—25 percent
Ni—25 percent

General Bell
Cu—80 percent
Sn—20 percent

Fire Engine Bells
Sn—20 percent
Ni—2 percent
Balance Cu

Large Bells
Cu—76 percent
Sn—24 percent

Railroad Signal Bells
Cu—60 percent
Zn—36 percent
Fe—4 percent

House Bells
Cu—76 percent
Sn—16 percent
Yellow Brass—8 per-cent

Gongs
Cu—82 percent
Sn—18 percent

Miscellaneous Metal Mixes

Slush Metal
Zinc..............92%
Aluminum........5%
Copper 3%

High Pressure Bronze
Copper 84%
Tin 7.5%
Lead 5.5%
Zinc......................3%

Phosphor Bronze
#1 copper wire 90%
Phosphor Tin.........10%

Aluminum Bronze
Copper 90%
Iron 05%
Aluminum...............9%
Manganese Copper...05%

Oriental Bronze
Copper 84%
Lead 10%
Tin.....................5%
Zinc....................1%

Index

A

Alloy forms, 135
Alloy, copper casting, 185
Alloy, nickel-silver, 182
Alloys, 78-79, 120, 154, 164, 177
Alloys, melting, 128
Alloys, red-metal, 181
Alloys, silicon copper, 153
Aluminum bronze, 167, 174
Aluminum, 65, 130
Antique dealers, 218

B

Beryllium bronze, 183
Brass and bronze mixes, 1, 241
Brass applications, 12
Brass fluxes, 241
Brass properties, 11
Brass, deoxidizing leaded red, 65
Brass, high-strength yellow, 74
Brass, leaded yellow, 67
Brass, melting, 52, 71, 161
Brass, red, 8, 43
Brass, semi-red, 51, 65
Brass, silicon, 150
Brass, yellow, 9, 67-68
Bronze or brass, 1
Bronze, aluminum, 167, 174
Bronze, beryllium, 183
Bronze, flaring manganese, 116
Bronze, heat treating aluminum, 176
Bronze, high-leaded tin, 131
Bronze, manganese, 77, 112
Bronze, melting, 115
Bronze, phosphor, 163
Bronze, phosphorized, 166
Bronze, silicon, 150
Bronzes, tin and leaded-tin, 119

C

Calcium boride, 35
Casting, gateless, 95
Casting, super precision, 224
Castings, art, 218
Castings, copper, 24
Castings, defective, 141
Chills, 110, 187

Choke, insufficient, 140
Copper content, 45
Copper ingots, 14
Copper molding sands, 16
Copper, casting, 13
Copper, gates and risers, 41
Copper, melting, 23
Copper, pouring, 36
Copper, test for gassy, 29
Copper-based alloys, 2
Core practice, 50, 70, 111, 128, 146, 161, 179
Core, skimmer, 91
Cores, 186
Coring, 169
Cover fluxes, 59
Crucible melting, 26
Crucible, 61, 116, 180
Cupola melting, 57
Cupola, 58
Cupro-nickels, 181

D

Dentrite, 121
Dentritic, 121
Deoxidizing and fluxing, 27, 57, 71, 129, 147, 180
Dross, 90, 93

E

Eutectic mixture, 169
Eutectic point, 169

F

Facing sand, 42
Facing, 48, 128, 136, 155, 179
Ferrous and nonferrous, 217
Fluxes, cover, 59
Fluxes, reducing, 59
Fluxing and deoxidizing, 57, 71, 129, 147, 162, 188
Foundry manipulation, 46
Foundry practice, 44, 67, 119, 134, 154, 177, 186
Foundry, clean, 225
Foundry, the non-foundry, 215
Furnace, arc rocking, 55
Furnace, rotary, 54

G

Gate construction, 92
Gate riser, 210
Gate, baffle-box, 208
Gate, bottom, 195
Gate, bottom-hub, 209
Gate, bow, 203
Gate, branch, 197
Gate, cap-core, 209
Gate, finder, 193
Gate, French, 200
Gate, horn, 192
Gate, horseshoe, 191
Gate, match-plate, 199
Gate, pencil, 192
Gate, reservoir, 201
Gate, ring, 194
Gate, riser whirl, 205
Gate, riser, 191
Gate, saxophone, 202
Gate, shrink-bob, 198
Gate, single-choke, 196
Gate, skimmer-sprue, 198
Gate, splash core, 195
Gate, step, 199
Gate, strainer-core, 197
Gate, through core, 202
Gate, umbrella, 190
Gate, wedge, 193
Gate, whirl, 195
Gates and risers for copper, 41
Gates and risers, 49, 70, 81, 128, 137, 155, 179, 189
Gating and risering, 140
Gating system, alternate, 98
Gating system, high-pressure, 159
Gating, 89
Gating, common errors in, 107
Grain distribution, 125

H

Heat treatment, 169

I

Impurities, 37, 72, 117, 130, 149, 180
Iron, 130
Isomorphism, 169
Isomorphous, 169

L

Ladle, alloying in the, 57

M

Machinability, 37, 187
Magnesium, 66

Matrix, 121
Molding sand, 47, 69, 80, 122, 136, 222

N

Nickel silvers, 182
Nozzling, 141

P

Phosphide migration, 164
Phosphorus, 28, 130
Precipitation hardening, 185

R

Red brass properties, 44
Red brass, 8, 43
Red metal mixes, 45
Red-metal alloys, 181
Riser connections, 143
Riser design, 144, 156
Riser gate, 191
Riser removal, 160
Riser, purpose of a, 95
Risering and gating, 140
Risers and gates, 49, 70, 81, 128, 137, 155, 179, 189
Risers, blind, 145
Rotary furnace, 54

S

Sand graph, typical, 125
Sand mixes, 240
Sand screening, 124
Sand, 142
Sand, copper molding, 16
Sand, dry, 18, 137
Sand, facing, 42
Sand, green, 21, 30
Sand, molding, 47, 69, 80, 122, 136, 155, 178, 222
Sand, tin bronze, 123
Saturated solution, 173
Silicon, 65, 130
Solid solution, 121
Sulfur, 65, 130

T

Test mold, 34
Test pattern, 32
Testing and degassing procedure, 33
Tools, lathe, 39

W

Weldability, 188
Worm gear, 96-99